3D Kinematics

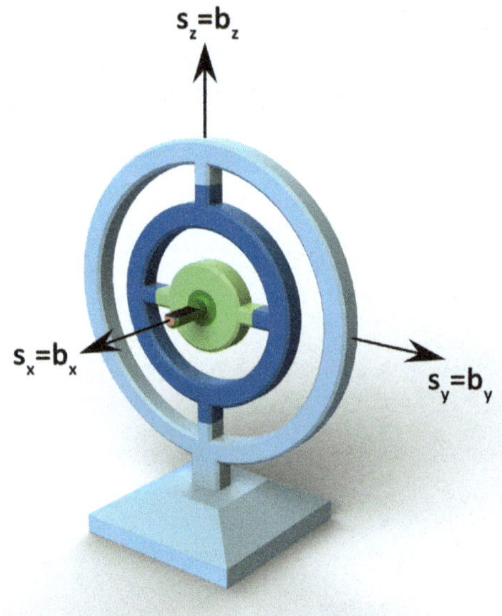

$s_z=b_z$
$s_x=b_x$
$s_y=b_y$

Thomas Haslwanter

3D Kinematics

Springer

Thomas Haslwanter
School of Medical Engineering and Applied
Social Sciences
University of Applied Sciences Upper
Austria
Linz, Upper Austria
Austria

Additional material to this book can be downloaded from http://extras.springer.com.

ISBN 978-3-319-75276-1 ISBN 978-3-319-75277-8 (eBook)
https://doi.org/10.1007/978-3-319-75277-8

Library of Congress Control Number: 2018933478

This Springer imprint is published by the registered company Springer Nature Switzerland AG
The registered company address is: Gewerbestrasse 11, 6330 Cham, Switzerland

This book is dedicated to Klaus Hepp, without whom I would have never worked in 3D kinematics, and to Ian Curthoys, who gave me the opportunity to apply it to image processing and to explain the underlying mathematics to others.

Preface

This book presents an introduction to the measurement and analysis of general movements in three-dimensional (3-D) space, i.e., movements containing a rotation and a translation component. It only deals with the "kinematics" of the movements, i.e., the description of the spatial position and orientation of objects; the body "dynamics", i.e., the description of the forces involved in the movements, are not dealt with here. I have been explaining 3-D kinematics to others almost as long as I can remember. First as a post-doc, when I was invited to write a review article on the mathematics of 3-D eye movements (Haslwanter 1995). Then to doctors, when we developed a computer program to simulate the effects of eye muscle surgeries on eye movements (Haslwanter et al. 2005). And more recently to students, while introducing them to the analysis of 3-D movement recordings. As a result of this background, the measurement techniques and the examples presented in this book are taken primarily from human movement analysis. But the same principles apply to calculations regarding 3-D orientation also in other areas, including aeronautics, astronomy, and computer vision.

Only a relatively small group of people have become comfortable with the sometimes complex formalisms required to describe orientation in space, and many researchers without a strong mathematical background have been deterred by the mathematical structures involved, e.g., quaternions or Euler angles. The geometrical background is presented here in such a way as to support the reader in developing a basic intuitive understanding of 3-D rotations. Although most relevant formulas are discussed, mathematical proofs published elsewhere have been largely omitted.

Since the most common techniques to record movements are measurements with inertial sensors and optical measurement systems, the book includes examples of movement recordings with these two types of sensors. I believe that especially the latter one might be interesting to a wider audience, since today every smartphone comes with built-in inertial sensors.

Many scientific researchers use either Python or Matlab for their data analysis, so implementations for most analysis steps are included as program packages for these two languages. I firmly believe that knowledge should be open and accessible, so personally I favor Python, as it is free and powerful, and has toolboxes for most common scientific problems. So for the code examples in this book I chose Python.

How to Use This Book

My most important advice is: sit down, read through the book, and try to do the calculations and exercises yourself! Only *reading* the book is not sufficient to gain understanding. Just as one will never learn how to play the piano by going to concerts, one also has to sit down, write down the algorithms, and program them in order to learn them! The software provided with this book should support getting results quickly.

Chapter 1 introduces the basic conventions and a few hands-on examples of working with vectors and matrices. For readers who are looking for a basic reminder of the most important concepts required from linear algebra and trigonometry, a short brush-up is given in Appendix A. In case of doubt, please go through it carefully, to make sure that your understanding of basic linear algebra, trigonometry, and numerical computation is solid enough to make it through the rest of the book.

Chapter 2 describes the main measurement techniques which can be used to record 3-D position and orientation.

Chapter 3 introduces rotation matrices for the description of the 3-D orientation of objects. The practical meaning of rotation matrices is explained, and examples are given to show their practical applications.

Chapter 4 introduces quaternions, another, more efficient way to characterize 3-D orientation. It also gives a short description of Gibbs vectors and Euler vectors, yet other ways to describe 3-D rotations.

Chapter 5 describes the algorithms to determine linear and angular velocity, with a focus on the connection between angular orientation and angular velocity.

Chapter 6 explicitly describes how 3-D position and 3-D velocity can be analyzed, based on data from either optical marker-based systems or from inertial measurement units (IMUs).

Chapter 7 presents the concept of sensor integration, which is used in real-life applications to compensate for drifts, bias, and noise in sensor data. Kalman filters, the most common approach to solve this problems, are described in more detail.

Appendices contain:

- mathematical details and proofs that are too long for presentation in the main text,
- descriptions and examples of the programming modules accompanying the book,

- step-by-step instructions for human movement recordings with optical systems or with inertial sensors,
- solutions to the exercises at the end of some chapters, and
- a list of online resources, for getting started as well as for further reading.

Acknowledgements

I want to thank Cornelius Willers, for highly valuable input on content and style. He provided the "outside view" for which I am really grateful. He also substantially improved my writing style, and supplied me with a practical application example that improved the manuscript. Thomas Pfandl has generated and rendered the 3-D images of gimbals and planes in this book, something I could not have done without him; he also checked some of the algorithms presented here. Andreas Kranzl has reviewed the chapter on measurements with optical marker-based systems, and provided helpful feedback based on his extensive experience with those systems. And my wife Jean has not only taken care of a loving and supportive environment but also helped me to improve the content and numerous illustrations.

Linz, Austria Thomas Haslwanter

Abbreviations

2-D	Two-dimensional
3-D	Three-dimensional
AHRS	Attitude and heading reference system
COM	Center of mass
CPU	Central processing unit
CRT	Cathode ray tube
CS	Coordinate system
CT	Computer tomography
DCM	Direction cosine matrix
DOF	Degrees of freedom
EMF	Electromotive force
GPS	Global positioning system
GPU	Graphical processing unit
IDE	Integrated development environment
IMU	Inertial measurement unit
IS	Inertial system
LOS	Line-of-sight
MARG	Sensor type providing magnetic, angular rate, and gravity information
MEMS	Microelectromechanical systems
ORS	Optical recording system
PC	Personal computer
SAW	Surface acoustic wave
SCS	Space-fixed coordinate system
SO3	3-dimensional special orthogonal group
SU2	2-dimensional special unitary group

Contents

1 Introduction 1
1.1 Recording Movement and Orientation 2
1.2 Conventions and Basics 2
1.2.1 Notation 3
1.2.2 Coordinate Systems 3
1.3 Software Packages 5
1.3.1 Python Package *scikit-kinematics* 5
1.3.2 Matlab 3-D *Kinematics Toolbox* 6
1.3.3 Source Code for Python and Matlab 7
1.4 Warm-Up Exercises 8

2 Measurement Techniques 11
2.1 Marker-Based Measurements 11
2.1.1 Image Formation 13
2.2 Sensor-Based Measurements 16
2.2.1 Overview 16
2.2.2 Linear Accelerometers 17
2.2.3 Gyroscopes 20
2.2.4 Ultrasound Sensors—Trilateration 22
2.2.5 Magnetic Field Sensors 23

3 Rotation Matrices 29
3.1 Introduction 29
3.2 Rotations in a Plane 30
3.2.1 Rotation in Cartesian Coordinates 31
3.2.2 Rotation in Polar Coordinates 32
3.2.3 Application: Orienting an Object in a Plane 33
3.3 Rotations About Coordinate Axes in 3-D 34
3.3.1 3-D Rotations About Coordinate Axes 34
3.3.2 Rotations of Objects Versus Rotations of Coordinate Systems 37

3.4 Combined Rotations 38
3.4.1 3-D Orientation with Sequential Rotations 40
3.4.2 Gimbal Lock 45
3.5 Homogeneous Coordinates 46
3.5.1 Definition 46
3.6 Applications 48
3.6.1 Two DOF—Targeting an Object in 3-D 48
3.6.2 Two DOF—Projection onto a Flat Surface 49
3.6.3 Three DOF—3-D Orientation Measurements with Search Coils 51
3.6.4 Nested or Cascaded 3-D Rotation Sequences 52
3.6.5 Camera Images 54
3.7 Exercises 54

4 Quaternions and Gibbs Vectors 57
4.1 Representing Rotations by Vectors 57
4.2 Axis-Angle Euler Vectors 59
4.3 Quaternions 59
4.3.1 Background 59
4.3.2 Quaternion Properties 60
4.3.3 Interpretation of Quaternions 61
4.3.4 Unit Quaternions 61
4.4 Gibbs Vectors 64
4.4.1 Properties of Gibbs Vectors 64
4.4.2 Cascaded Rotations with Gibbs Vectors 66
4.4.3 Gibbs Vectors and Their Relation to Quaternions 66
4.5 Applications 66
4.5.1 Targeting an Object in 3-D: Quaternion Approach 66
4.5.2 Orientation of 3-D Acceleration Sensor 68
4.5.3 Calculating Orientation of a Camera on a Moving Object 70
4.5.4 Object-Oriented Implementation of Quaternions 72

5 Velocities in 3-D Space 75
5.1 Equations of Motion 75
5.2 Linear Velocity 76
5.3 Angular Velocity 79
5.3.1 Calculating Angular Velocity from Orientation 79
5.3.2 Calculating Orientation from Angular Velocity 82

6 Analysis of 3-D Movement Recordings 85
6.1 Position and Orientation from Optical Sensors 85
6.1.1 Recording 3-D Markers 85
6.1.2 Orientation in Space 87
6.1.3 Position in Space 88

6.1.4 Velocity and Acceleration 89
6.1.5 Transformation from Camera- to Space-Coordinates 89
6.1.6 Position 90
6.2 Position and Orientation from Inertial Sensors 91
6.2.1 Orientation in Space 91
6.2.2 Position in Space 93
6.3 Applications: Gait Analysis 95
6.4 Exercises 97

7 Multi-sensor Integration 99
7.1 Working with Uncertain Data 100
7.1.1 Uncertain Data in One Dimension 100
7.1.2 Uncertain Data in Multiple Dimensions 102
7.2 Kalman Filter 105
7.2.1 Idea Behind Kalman Filters 105
7.2.2 State Predictions 107
7.2.3 Measurements and Kalman Equations 109
7.2.4 Kalman Filters with Quaternions 111
7.3 Complementary Filters 111
7.3.1 Gradient Descent Approach 112

Appendix A: Appendix—Mathematics 115

Appendix B: Practical Applications: Denavit-Hartenberg Transformations 129

Appendix C: Python and Matlab Programs 133

Appendix D: Human Movement Recordings—Practical Tips 155

Appendix E: Exercise Solutions 161

Appendix F: Glossary 179

Appendix G: Online Resources 183

References 185

Index 189

Chapter 1
Introduction

Performing an everyday movement, such as reaching for a cup of tea, is so natural and intuitive to us that it seems to be trivial. But when we try to understand how this movement is performed, or when we try to follow or imitate such a movement, for example, with a robotic arm, it quickly becomes obvious that even such seemingly trivial acts are based on a complex interaction of the relative three-dimensional (3-D) upper body, arm, and finger orientations. Similarly, looking at the face of an approaching friend while walking down the street does not seem to be much of an achievement. But talk to an engineer who has tried to keep a camera on a moving platform oriented such that it keeps focussing on another moving target, and you realize that working with objects moving in 3-D space entails many challenges, especially mathematical and geometric ones.

Surprisingly, little literature exists that provides a researcher or engineer who wants to work on this type of phenomena with an introduction into the area. On the contrary, most articles or books focus on one selected way to characterize a 3-D movement, but do not elaborate on alternative ways to describe it. For example, my own physics education gave me a (confusing) introduction to "Euler angles" or the "special unitary group of complex 2×2 matrices", but never showed how to work with them in practice, and did not mention alternative descriptions of spatial orientation, such as quaternions.

This book tries to fill this gap. It will provide an overview of common ways to characterize movement in 3-D space. In particular, it will provide an introduction to the different methods that are commonly used to record and analyze human movements, be it for medical applications (such as gait analysis), scientific uses (such as biomechanical investigations), or for recreational activities (such as the movement analysis with the sensors built into current smartphones). But it should also be able to provide programmers working in computer graphics with the necessary background to choose the optimal algorithms for their kinematic tasks at hand.

To my knowledge, this book is the first one that not only describes the mathematics of 3-D kinematics but also provides full programming toolboxes (in Python and

T. Haslwanter, *3D Kinematics*, https://doi.org/10.1007/978-3-319-75277-8_1

in Matlab), allowing the reader to focus on the understanding and not on "trivial" programming details. The Python package *scikit-kinematics*,[1] as well as a corresponding Matlab *Kinematics Toolbox*,[2] contain the algorithms for simulating 3-D movements, and for importing and analyzing data from different 3-D recording systems. Code listings and the solutions to the exercises can be found on the website accompanying this book.[3]

1.1 Recording Movement and Orientation

Determination and characterization of orientation and movement in space can provide valuable information for numerous applications:

- Smartphones use such measurements to decide whether the display should be in portrait or landscape mode.
- Fitness trackers, such as *Jawbone* or *Fitbit*, use this information to estimate and quantify the amount of daily movement activities.
- Airbags in cars are triggered by movement sensors.
- In neurology, otorhinolaryngology, and ophthalmology, movement recordings are used for the diagnoses of medical conditions.
- Autopilot applications in planes and autonomous vehicles require movement information for their actions.
- Modern prosthetic devices include movement sensors, to control built-in motors and to regulate the mechanical properties of modern prostheses.

Simple approaches are often sufficient for two-dimensional (2-D) measurements. A simple protractor is sufficient to find the angles between upper body, upper leg, and lower leg from a photography of a runner. And a goniometer can quickly indicate the angle between two objects or shafts.

However, to uniquely characterize the movement of an object in 3-D space, the measurements are more involved and six parameters are required. For recording of 3-D position and orientation, which together are sometimes referred to as pose, two approaches can be taken. First, three or more parts of an object can be marked. Tracking the movement of those markers in 3-D space provides information about the movement of the object. And second, if the object is solid, a sensor can be attached to the object. The signals from this sensor can then be used to find the position and orientation of the sensor, and thus of the object.

1.2 Conventions and Basics

Movements in 3-D space consist of translations as well as rotations. To describe them, the following conventions will be used.

---

[1] https://github.com/thomas-haslwanter/scikit-kinematics.

[2] https://github.com/thomas-haslwanter/kinematics_toolbox.git.

[3] https://github.com/thomas-haslwanter/3D_Kinematics.

1.2.1 Notation

- Axes indexing starts at 0, (0, 1, 2) and corresponds to the (x, y, z) axes, respectively.
- Scalars are indicated by plain letters (e.g., a).
- Column vectors are written with bold lowercase letters (e.g., $\mathbf{r}$) or in round brackets, and the components of 3-D coordinate systems are labeled (x, y, z):

$$\mathbf{r} = \begin{pmatrix} r_x \\ r_y \\ r_z \end{pmatrix}.$$

(The only exception are the electrical field **E** and the magnetic field **B**, which by convention are written in uppercase Sect. 2.2.5). However, it should be clear from the context that they are vectors.)
- The length or "norm" of a vector is indicated by the same name but in plain style

$$|\mathbf{r}| = \sqrt{\sum_i r_i^2} = r.$$

- Matrices are written with bold uppercase letters (e.g., $\boldsymbol{R}$) or in square brackets.

$$\boldsymbol{R} = \begin{bmatrix} R_{xx} & R_{xy} & R_{xz} \\ R_{yx} & R_{yy} & R_{yz} \\ R_{zx} & R_{zy} & R_{zz} \end{bmatrix}.$$

- Vector and matrix elements are written in plain style, with indices denoted by subscripts (e.g., r_x; R_{yz}).
- Multiplications with a scalar are denoted by $*$ (e.g., $\tan(\theta/2) * \mathbf{n}$).
- Scalar–vector products and matrix multiplications are denoted by $\cdot$ (e.g., $\mathbf{p} \cdot \mathbf{q}$).
- Vector cross products are denoted by $\times$ (e.g., $\mathbf{p} \times \mathbf{q}$).
- Quaternions are denoted with bold italics and tilde (e.g., $\tilde{\boldsymbol{r}}$).
- Products of quaternions or Gibbs vectors are denoted by $\circ$ (e.g., $\tilde{\boldsymbol{r}}_{\boldsymbol{p}} \circ \tilde{\boldsymbol{r}}_{\boldsymbol{q}}$).

1.2.2 Coordinate Systems

A frequent source of confusion is the choice of coordinate system. Unit vectors in the direction of the x-, y-, z-axes will be denoted with $\mathbf{n_x}$, $\mathbf{n_y}$, $\mathbf{n_z}$, respectively. The direction of $\mathbf{n_x}$ can be chosen freely. For example, it can point forward, left, or up.

Modern texts almost exclusively use right-handed coordinate systems (Fig. 1.1), but may attach different meanings to the three axes. For example, in image processing $\mathbf{n_x}$ is typically chosen pointing right and $\mathbf{n_y}$ pointing up so that the image plane is the (x, y)-plane. In aerospace engineering, $\mathbf{n_x}$ is pointing forward, $\mathbf{n_y}$ is chosen such that

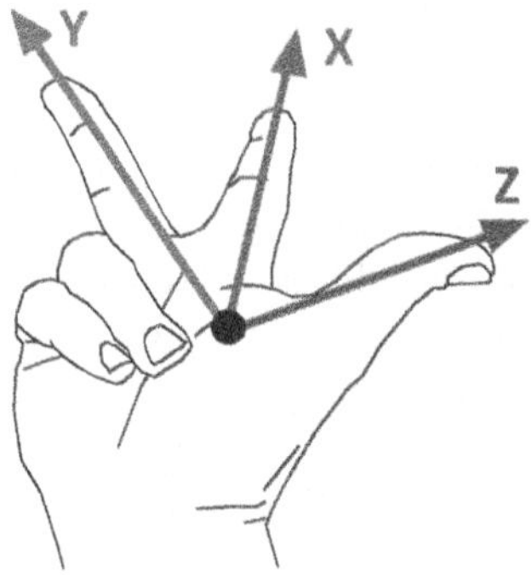

Fig. 1.1 Right-hand reminder for the direction of the positive coordinate axes. Remember where "x" is pointing to! (from Wikipedia, by R. Hewitt)

it points to the right, and $\mathbf{n_z}$ as a result is pointing down. With that convention, nose-up rotations of an airplane are "positive", the preferred choice in aeronautics. When used in navigation the axes order may denote East-South-Down or North-East-Down. And in human locomotion analysis $\mathbf{n_x}$ should point in the direction of progression, $\mathbf{n_y}$ upward, and $\mathbf{n_z}$ to the right (Wu and Cavanagh, 1995). But regardless of the specific choice, it is very important to make sure which coordinate system has been selected.

In this book, the default coordinate system will be a right-handed coordinate system with three orthogonal unit vectors. The coordinate system is chosen as it is commonly used in medical applications and movement analysis. It defines the axes as follows (Fig. 1.2):

- $\mathbf{n_x}$ pointing forward,
- $\mathbf{n_y}$ pointing to the left, so that the x, y-plane ($z = 0$) is horizontal, and
- $\mathbf{n_z}$ pointing up.

so that

$$\mathbf{n_x} \times \mathbf{n_y} = \mathbf{n_z} \,. \tag{1.1}$$

Wherever possible the axis labels *("x", "y", "z")* will be used to avoid labeling by numbers *("0", "1", "2")*, since some computer languages (like C or Python) start with 0, while others (like Matlab) start with 1.

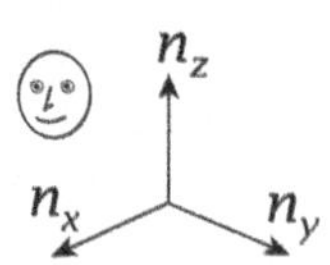

Fig. 1.2 Right-handed coordinate system

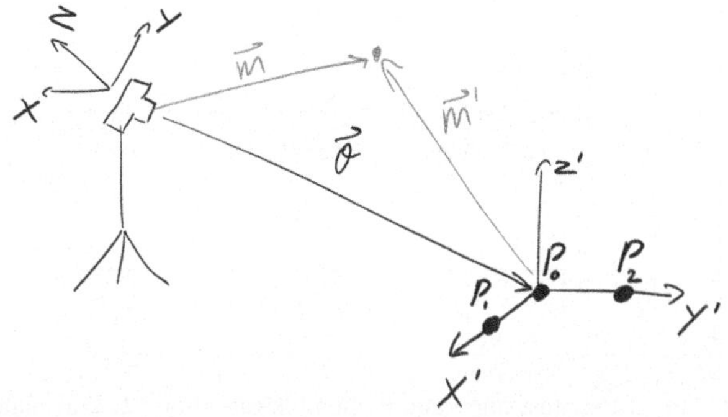

Fig. 1.3 For finding a correct mathematical solution to the individual problem at hand, informal sketches are invaluable! In most cases, the programming should be almost trivial, especially when using the software provided with this book. But 3-D kinematics is complex to visualize, and the help provided by simple sketches is hard to overestimate (Here, a sketch for a camera-based recording of an experimental setup, as will be used in Chap. 6.)

1.3 Software Packages

To facilitate and speed up the analysis of 3-D data, this book comes with libraries in Matlab and Python. These libraries provide frequently used functions for working with vectors, rotation matrices, and quaternions, and for the data analysis for measurements from inertial measurement units (IMUs) or from optical recording systems (e.g., Optotrak or Vicon) (Fig. 1.3).

The application examples in this book are presented in Python. The corresponding source code can be found on the web-page accompanying this book.[4] A list of the programs included is given in Appendix C.

1.3.1 Python Package scikit-kinematics

The Python core distribution contains only the essential features of a general programming language. For example, it does not even contain a package for working efficiently with vectors and matrices. These packages, and many more that are useful for scientific data analysis, can be installed most easily using so-called "Python distributions". Two recommendable Python distributions are

- *WinPython* for Windows only.
- *Anaconda* by Continuum, for Windows, Mac, and Linux.

Both distributions are freely available, and neither requires administrator rights. A list of links for the downloads of these distributions, as well as recommendations for getting started with Python for scientific applications, can be found in Appendix G.

The relationships between the basic scientific Python packages used by *scikit-kinematics* is shown in Fig. 1.4, as well as the role of *Jupyter* and *IPython* which are used for interactive data analysis.

---

[4]https://github.com/thomas-haslwanter/3D_Kinematics.

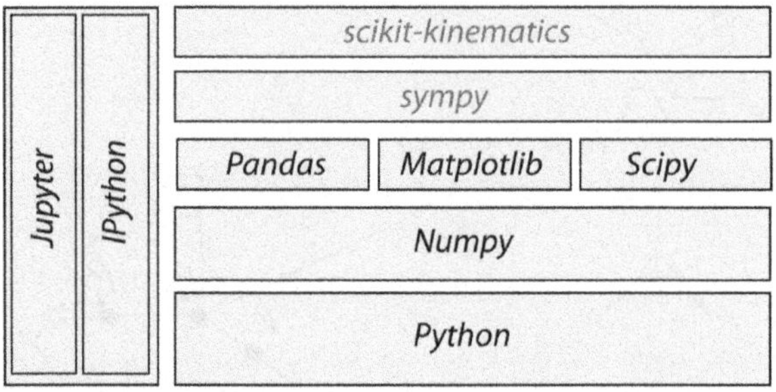

Fig. 1.4 The structure of the most important Python packages for 3-D kinematics. The standard scientific packages are written in black; more specialized packages are labeled in gray. `sympy` will be used here for working with symbolic matrices

The programs included in this book have been tested with Python 3.6.3 under Windows and Linux using the following package versions:

- *Jupyter 1.0.0* … Framework for interactive work.
- *IPython 6.2.1* … Python kernel for interactive work.
- *numpy 1.13.3* … For working with vectors and arrays.
- *scipy 1.0.1* … All the essential scientific algorithms, including those for basic statistics.
- *matplotlib 2.2.2* … The de-facto standard package for plotting and visualization.
- *pandas 0.22.0* … Adds "DataFrames", which are easy to use data structures, to Python.

Building on this basis, the Python package *scikit-kinematics* is intended to facilitate the development of programs for the analysis of spatial data. It can be downloaded from https://github.com/thomas-haslwanter/scikit-kinematics and is documented under http://work.thaslwanter.at/skinematics/html/. The easiest way to install it is by typing

```
pip install scikit-kinematics
```

on the command line. Updates can be performed with

```
pip install --upgrade --no-deps scikit-kinematics
```

In the Python applications, *scikit-kinematics* is for brevity referred to as `skinematics` (Fig. 1.5).

1.3.2 *Matlab 3-D* Kinematics Toolbox

Matlab is the 800-pound gorilla in the room when it comes to scientific computing. It has been around for a long time (I have used Matlab for more than 20 years) and is well established in many academic and industrial environments. In contrast to

Fig. 1.5 The scikit-kinematics logo

Python, which is a general programming language, Matlab is tailored to numerical applications. It is a fully developed integrated development environment (IDE) and has a wealth of "Toolboxes" available, which are extensions for dedicated programming applications.

The downsides of Matlab are that it is commercial, expensive for those outside an academic environment, and that—compared to Python—it is a rather old programming language. Matlab's object-oriented programming scheme is unwieldy and overly complex.

The 3-D Kinematics toolbox accompanying this book can be downloaded from the Matlab Kinematics Toolbox [5] and can be installed simply by opening the file `3D_Kinematics.mltbx` in Matlab. The toolbox files will then be copied to the correct locations in Matlab, and the corresponding search path added to the `MATLABPATH`.

1.3.3 Source Code for Python and Matlab

The Python package *scikit-kinematics* and the Matlab *Kinematic toolbox* are shared via https://github.com/thomas-haslwanter.

A frequent source of confusion is the difference between "git" and "github". *git* is a "version control program", whereas *github* is a website.

Version control programs (such as *git*), also known as revision control programs, allow tracking only the modifications, and storing previous versions of the source code under development. If the latest changes cause a new problem, it is then easy to compare them to earlier versions, and to restore the source code to a previous state. Git can be used locally, with very little overhead. And it can also be used to maintain and manage a remote backup copy of the code. While the real power of git lies in its features for collaboration, it is also powerful and works very smoothly for personal software development. git is well integrated into most Python IDEs, and in Matlab.

Under Windows *tortoisegit* (https://tortoisegit.org/) provides a very useful Windows shell interface for git. For example, in order to clone a repository (e.g., *scikit-*

[5] https://github.com/thomas-haslwanter/kinematics_toolbox.git.

kinematics or the *Kinematics Toolbox*) from github to a computer where tortoisegit is installed, one simply has to right click on the folder where one wants the repository to be installed, select `Git Clone ...`, and enter the repository name—and the whole repository will be cloned there. Done!

github is a website frequently used to share code. While one can download source code from there, it is much more efficient to use *git* for this task.

python **Code**: `C1_examples_vectors.py`: Example of working with vectors. (p.133)

1.4 Warm-Up Exercises

This first batch of examples is intended as a reminder of the basic principles of geometry, trigonometry, and numerical analysis. Solutions to these exercises are provided in Appendix E.

Exercise 1.1: A Simple Linear Movement
An accelerometer moving sinusoidally along a single axis indicates an output (Fig. 1.6)

$$acc(t) = amp * \sin(\omega t)\,. \tag{1.2}$$

Knowing the initial conditions $vel(t = 0)$ and $pos(t = 0)$, it is possible to determine the movement of the accelerometer in space. Please try to do that analytically.

Exercise 1.2: Find the Cat
Take the image in Fig. 1.7, showing me and my three-legged cat Felix, and the following additional information:

- The coordinate center is defined as the center position on the ground between my legs.
- The Ikea shelf behind me has a height of 1.24 m.

Try to answer the following question, using only a simple drafting triangle:

"What are the coordinates of the cat (e.g., the center between the cat eyes) in a space-fixed coordinate system, defined as (x, y, z) pointing forward, left, and up, respectively?"

List the required steps, as well as all the assumptions made. Make a sketch of the geometry of the problem and write down the equations that would be needed to solve it.

Fig. 1.6 Sinusoidal movement along one dimension

Fig. 1.7 Me and my cat Felix

Exercise 1.3: Simple Pendulum

At first sight, a pendulum executes a deceptively simple motion. For example, for small swings the movement is nicely sinusoidal.

Assume that a pendulum with a length of $r = 0.2\,\mathrm{m}$ and a mass of $m = 0.5\,\mathrm{kg}$, deflected by an angle of θ_0, is released at t=0. Find the position of the pendulum for times $0\,\mathrm{s} <= t <= 10\,\mathrm{s}$, with a $\Delta t = 1\,\mathrm{ms}$, for initial deflections of $5°$ and of $70°$ (Fig. 1.8).

The movement of a pendulum can be simulated using *Newton's second law*

$$L = I * \frac{\mathrm{d}^2\theta(t)}{\mathrm{d}t^2}, \tag{1.3}$$

where L is the torque and I is the moment of inertia. For a pendulum, the moment of inertia is $I = m * r^2$. And the torque L is given by $L = r * F$, where F is the tangential force acting on the pendulum. The equations for deflection θ and angular velocity $\omega = \frac{\mathrm{d}\theta}{\mathrm{d}t}$ can be solved iteratively:

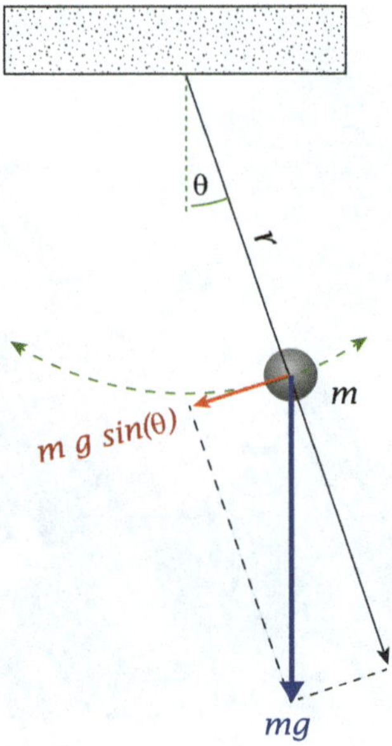

Fig. 1.8 A "simple" pendulum

$$\theta(t_{n+1}) = \theta(t_n) + \omega(t_n) * \Delta t \tag{1.4}$$

$$\begin{aligned}\omega(t_{n+1}) &= \omega(t_n) + \left.\frac{d^2\theta}{dt^2}\right|_{t_{n+1}} * \Delta t \\ &= \omega(t_n) + \left.\frac{L}{I}\right|_{t_{n+1}} * \Delta t \end{aligned} \tag{1.5}$$

Hints:

- First, write down the implementation of the equations for $\theta(t_i)$ and $\omega(t_i)$.
- Note that to improve the stability of the solution, the *Euler–Cromer method* is used in Eq. (1.5): this means that for the acceleration term $L(t_{n+1})$ is used, not $L(t_n)$!

Exercise 1.4: Not-so-simple Pendulum

If Exercise 3 is not challenging for you, try to answer the following question:

If the mass at the end of the pendulum is replaced by an accelerometer, what will the output of that accelerometer be when we let go of the pendulum, from an initial deflection of 10°?

The answer to this question is surprising, and surprisingly difficult to write down. Do not worry if you have difficulties solving this problem now, but give it a try again after having completed Chap. 6.

Chapter 2
Measurement Techniques

This chapter will give an overview of the measurement principles behind the optical- and sensor-based methods that are most commonly used to record 3-D movements. Details for human motion capture can be found in the recent Handbook of Human Motion [Müller et al., 2018].

2.1 Marker-Based Measurements

Essentially, every identifiable feature of an object can be taken as a marker or interest point:

- In image processing, corners, defined as the intersection of two edges, are often used as markers, in order to track objects or to compare images for similarity. (The subsequent mapping of one image to another is called "image registration".) A number of algorithms exist for corner detection, such as the Harris–Stephens algorithm (Harris and Stephens 1988).
- Particles or bubbles in gases or fluids (Hassan and Canaan 1991), or gold beads inserted into muscles (Miller et al. 2003), can be tracked and have been used to determine movement and deformation of elastic, viscous, liquid, and gaseous materials.
- Combined information of images and depth sensors can be used to estimate joint locations, an approach used very successfully in the *Microsoft Kinect* (Kar 2010).
- Passive markers, which can be anything from a simple "*X*" painted onto the item to be tracked or small 3-D spheres which selectively reflect visible or infrared light, can also be used to track well-defined parts of an object of interest. In the area of life sciences, the *Vicon* system has become almost a gold standard (Fig. 2.1). Passive marker systems are easier to handle than active marker systems (see below), because the markers do not require any cables.

T. Haslwanter, *3D Kinematics*, https://doi.org/10.1007/978-3-319-75277-8_2

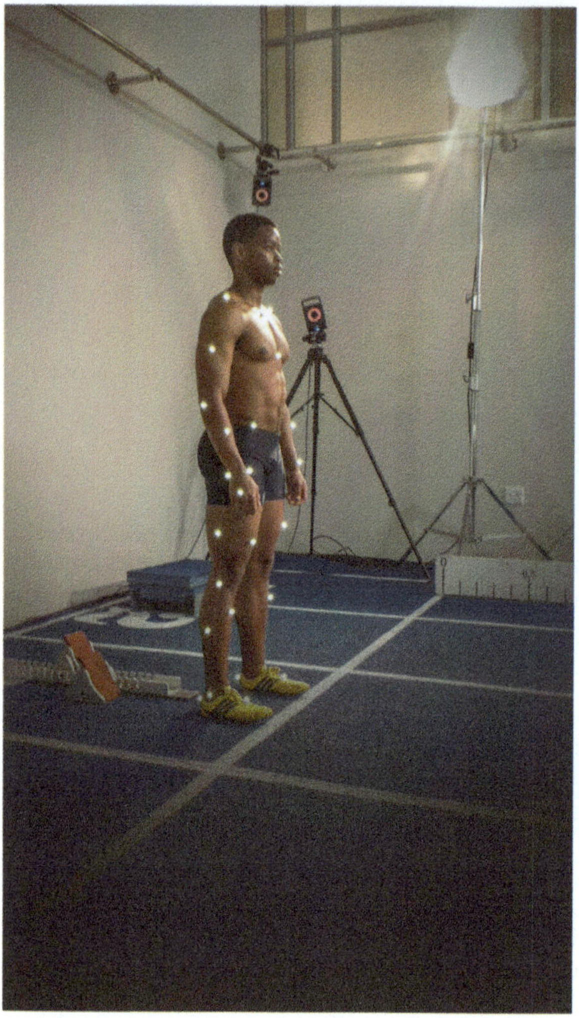

Fig. 2.1 A *Vicon* system for human motion capture. Passive markers, which are small, reflective spheres, are first attached to interesting points of the human body. Using an array or infrared cameras, the location of those markers in space can then be reconstructed (With kind permission from *Vicon*)

- Active markers, which emit light or sound signals at known points in time, can be attached to subjects or objects, and can be tracked. Active markers can be *sound markers* or *optical markers*. Sound markers are located based on the runtimes between the pulsed sound signal and three or more microphones. In contrast, optical markers are typically located with video cameras (Fig. 2.2). Thereby, high-speed line-scan cameras can be used instead of 2-D imaging sensors, allowing higher sampling frequencies. The advantage of active marker systems is

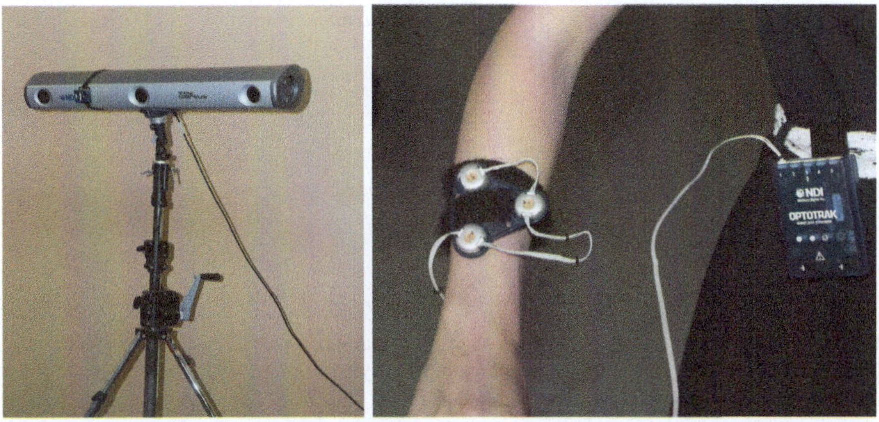

Fig. 2.2 Left: The Optotrak sensor with three cameras, mounted on a tripod. Right: The corresponding active markers (here three combined), with the communication unit to communicate with the sensors

that the markers can always be uniquely identified, even if they were obscured intermittently.

The algorithms to find the location of a marker in 3-D space depend on the type of marker. *Trilateration*, i.e., the process of determining absolute or relative locations of points by measurement of distances, is used for sound markers. This topic will be covered briefly in Sect. 2.2.4. When signals are obtained by cameras the measurement technique is called "stereophotogrammetry", and image processing is typically used to find the location of a point of interest in a 2-D image. The underlying principles of image formation are described below. The reconstruction of a 3-D location by combining information from two or more cameras is not covered in this book.

2.1.1 Image Formation

Many measurements of 3-D objects are performed with optical systems, which project 3-D data into a 2-D image plane. Typically, the image plane is assigned the (x, y)-coordinates, with the positive x-axis pointing right (along image columns), and the positive y-axis pointing up or down (along image rows), thus placing the origin in the bottom-left or in the top-left corner, respectively.

Projections describe the mapping of the 3-D object space into the 2-D image space.

Parallel Projection

The simplest transformation from a 3-D object into the image plane is the "parallel projection", for example, the projection of the pupil center of an eye into an image plane (Fig. 2.3). The location of the pupil center in 3-D is given by

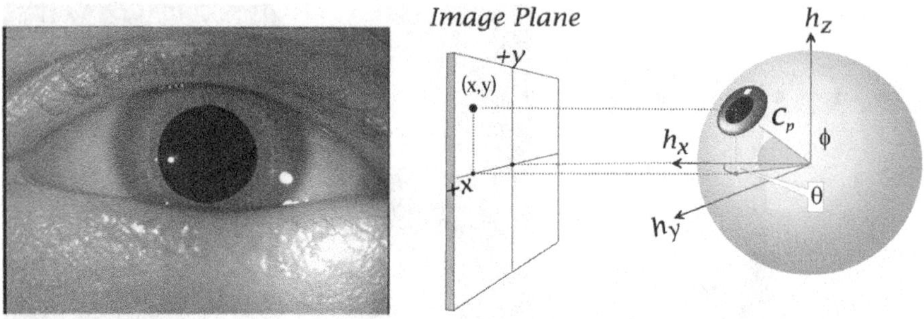

Fig. 2.3 Parallel projection from 3-D into a plane. Note that the coordinate system of the image plane (x, y) is different from the head/space-fixed coordinate system (h_x, h_y, h_z). θ and ϕ correspond to the *nautical sequence* of rotations (see section 3.4.1)

$$\mathbf{c_p} = \begin{pmatrix} x \\ y \\ z \end{pmatrix}_{\text{SpaceFixed}} . \tag{2.1}$$

With the setup in Fig. 2.3, the space-fixed (y, z)-plane is parallel to the image plane. Visual inspection of Fig. 2.3 shows that

$$\begin{pmatrix} x \\ y \end{pmatrix}_{\text{ImagePlane}} = \begin{pmatrix} y \\ z \end{pmatrix}_{\text{SpaceFixed}} . \tag{2.2}$$

Generalization: The general case is a bit more complicated. It involves the "rotation matrices" introduced in Chap. 3. Here, the rotation matrix **R** describing the rotation from the imaging system to the space-fixed system is given by (see Chap. 3)

$$\mathbf{R} = \begin{bmatrix} 0 & 1 & 0 \\ 0 & 0 & 1 \\ 1 & 0 & 0 \end{bmatrix} . \tag{2.3}$$

And the equations for the transformation of the coordinates can be derived in the same way as Eq. (6.15) in Sect. 6.1.6. The result is presented here without proof: If **c** is the point under consideration, $\mathbf{p}_{CS}$ indicates the center of the space-fixed coordinate system expressed relative to the imaging system, and **R** the rotation matrix describing the orientation of the space-fixed coordinate system relative to the imaging system, then the position of the point with respect to the imaging system is

$$\mathbf{c}^{\text{image}} = \mathbf{R} \cdot \mathbf{c}^{\text{space}} + \mathbf{p}_{CS} , \tag{2.4}$$

and the projection into the image plane is given by

$$\mathbf{c}^{\text{image}}_{\text{projected}} = \mathbf{c}^{\text{image}} - (\mathbf{c}^{\text{image}} \cdot \mathbf{n_z}) * \mathbf{n_z} , \tag{2.5}$$

where $\mathbf{n_z}$ is the unit vector along the z-axis of the imaging system.

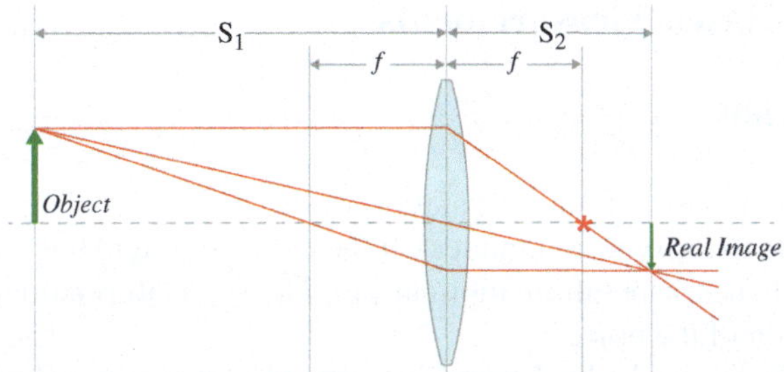

Fig. 2.4 Central projection through a thin lens. The red "*" indicates the focal point of the lens

For practical applications, 2-D projections are also implemented in the `scikit-kinematic` function `vector.project`, when it is called with the option `projection_type='2D'`.

Central Projection

The image of an object is typically obtained using optics. In the simplest case, the optics consist of a single lens. If the object is very close to the camera, a parallel projection no longer describes the imaging process accurately. The projection of an object through a lens into the image plane of a camera can be approximated more accurately by a "central projection" (Fig. 2.4).

For a central projection, object distance S_1, focal length f, and image distance S_2 are related by the *paraxial*[1] *thin lens equation*

$$\frac{1}{S_1} + \frac{1}{S_2} = \frac{1}{f}. \tag{2.6}$$

For many practical applications, the objects are much further away than the focal length $S_1 >> f$. In that cases, the image is located close to the focal plane, and $S_2 \approx f$.

The paraxial central projection only holds for objects close to the optical axis. If objects are located at a significant distance from the optical axis, or for objects very close to the lens, the technique of "ray-tracing" has to be used to determine the accurate image of the object computationally.

Note: Images are typically acquired digitally with CCD or CMOS image sensors. As a result, the image values are expressed in pixels, not in millimeters or meters. If possible, also the image of a reference object, e.g., a ruler, should be acquired, to check that the assumed conversion factor from pixels to length is correct.

---

[1]In the paraxial approximation, Equations A.7 and A.8 are made, assuming small field angles and small image heights.

2.2 Sensor-Based Measurements

2.2.1 Overview

Information about position and orientation of an object can also be obtained by attaching sensors to the object of interest. If the object is a rigid body, and the sensor tightly attached to it, sensor information can provide direct information about position and/or orientation of the object.

Sensors can be screwed, glued, taped, or strapped to objects in order to measure the movement of the objects. But it is surprisingly difficult to measure the kinematics of a human body, to decide what exactly should be recorded, and where to firmly attach sensors. Consider the movement of the lower arm: the lower arm has two bones, radius and ulna, and those can move with respect to each other (especially with a rotation of the wrist); and the muscles can contract and thereby change their thickness, even when the bones are stationary. On top of that, the skin overlying the muscles adds an additional possible source of sensor movement. Measurements where markers were actually inserted into the bone (so-called "bone pins") have tried to quantify the errors obtained when using surface-mounted markers in biomechanics (Reinschmidt et al. 1997a, b). But for most applications, such bone pins are not practical.

Sensors such as accelerometers, gyroscopes, Hall sensors (for static magnetic fields), and induction sensors (for dynamic magnetic fields) are commonly used for kinematic recordings. Accelerometers and gyroscopes are referred to as "inertial sensors" (Fig. 2.5), since they provide a signal when the position or orientation of an object is moved with respect to an inertial system.

Sensors for static magnetic fields, like the local magnetic field of the earth, provide absolute orientation information, even without any movement of the object. Static magnetic field sensors use the Hall effect for signal generation (see Sect. 2.2.5). Induction sensors provide orientation information in oscillating magnetic fields, which have to be generated externally in order to obtain a signal.

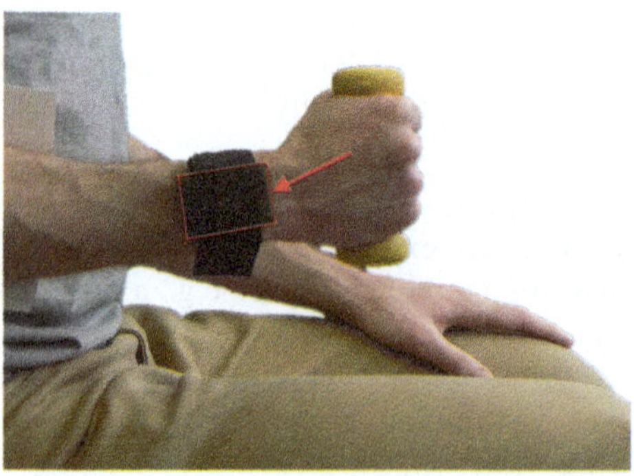

Fig. 2.5 Inertial sensor from *x-io Technologies* (red box/arrow) attached to the wrist during a physiotherapy exercise

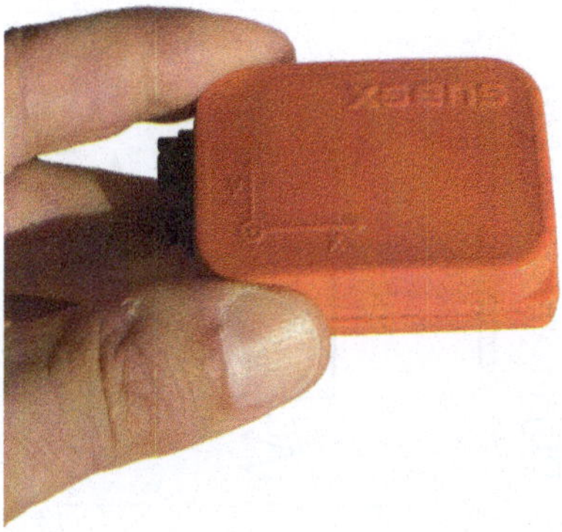

Fig. 2.6 The MTX sensor from Xsens measures linear acceleration, angular velocity, and the local magnetic field

A very good overview of "inertial measurement units" (IMUs) has been provided by (Woodman 2007), and a number of the following details have been taken from that article. The most common type of IMUs in biomedical applications are strapdown[2] systems, sensors that can be strapped to the body. They are typically produced with "microelectromechanical systems" (MEMS) technology, which enables the creation of very small electromechanical devices using semiconductor fabrication techniques. Figures 2.5 and 2.6 show examples of integrated devices containing three types of sensors:

- **Accelerometers** to measure linear accelerations,
- **Gyroscopes** to measure angular velocity, and
- **Magnetometers** to measure the 3-D orientation of the local magnetic field.

2.2.2 Linear Accelerometers

Accelerometers sense the "gravito-inertial force" (GIF), i.e., the sum of gravity and inertial forces caused by linear accelerations.

$$\mathbf{f}_{\mathrm{GIF}} = \mathbf{f}_{gravity} + \mathbf{f}_{LinAcc} = m * \left(\mathbf{g} + \frac{d^2\mathbf{x}}{dt^2}\right). \tag{2.7}$$

This gravito-inertial force can be measured either mechanically or with a solid-state device. Mechanical measurements can be performed based on a number of electromechanical effects. For example, a mechanical accelerometer can read out the displacement of a "proof mass", as shown in Fig. 2.7. Or it can be designed as a "Surface Acoustic Wave" (SAW) accelerometer (Fig. 2.8). A SAW accelerometer uses a cantilever beam resonating at a particular frequency. A mass is attached to one

[2] Strapdown sensors mean that the sensor is firmly fixed to the object being measured, and there are no gimbals or moving parts.

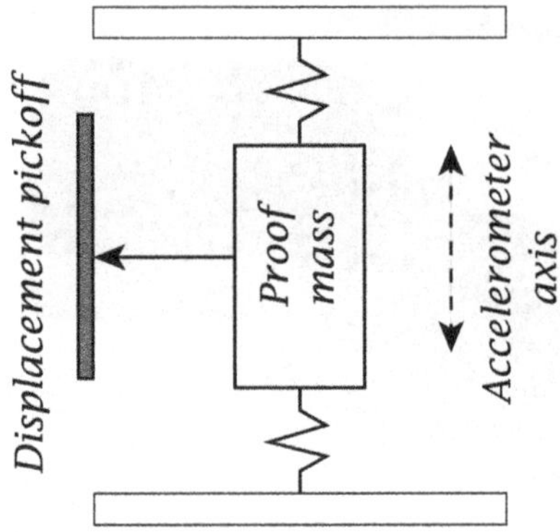

Fig. 2.7 A mechanical accelerometer

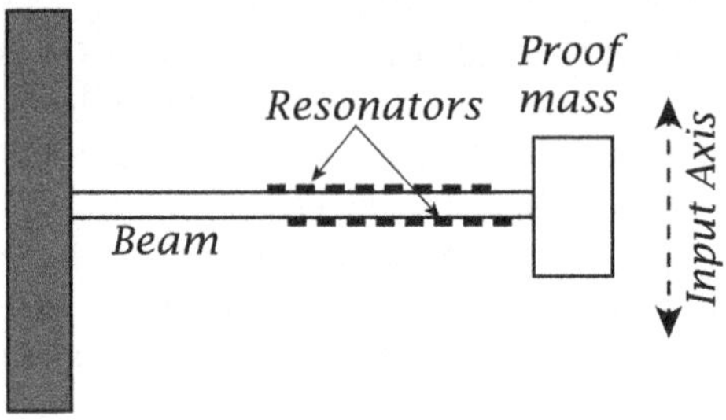

Fig. 2.8 Sketch of a surface acoustic wave (SAW) accelerometer

end of the beam which is free to move. The other end is rigidly attached to the case. When an acceleration is applied along the input axis, the beam bends. This causes the frequency of the surface acoustic wave to change proportionally to the applied strain. By measuring this change in frequency with identical patterns of surface acoustic wave delay lines or resonator electrodes on two opposing sides of the beam, the acceleration can be determined. This type of accelerometer is useful for high impact and vibration applications.

Sign of Gravitational Acceleration

The linear acceleration forces acting on a sensor are the sum of gravity and the linear accelerations elicited by movements in space [(Eq. (2.7)]. But since the sensor cannot distinguish between gravitational and inertial forces, everything is interpreted as "acceleration", irrespective of the cause of the force. (An extension of these ideas would lead to the theory of special relativity.)

Note that according to Newton's third law "*For every action, there is an equal and opposite reaction*", the direction of the *force* causing the movement and the corresponding sensed *acceleration* have the opposite direction (see Fig. 2.9)! For example, if you drive a powerful car and accelerate *forward*, you are pushed *backward* into the driver seat. So for a sensor lying stationary on the floor, the gravitational

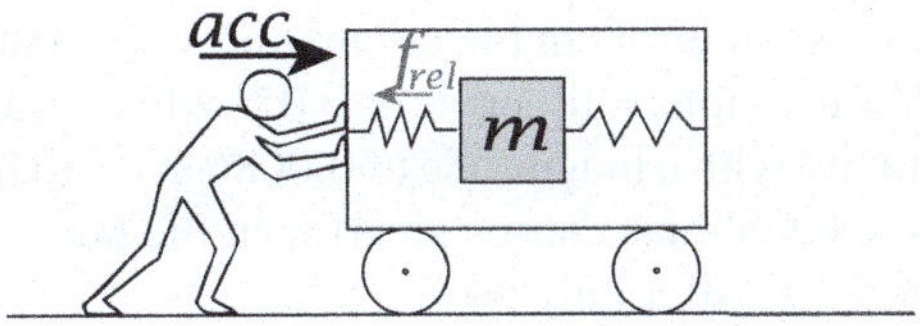

Fig. 2.9 Acceleration in one direction causes a relative force $\mathbf{f}_{rel}$, and a corresponding mass displacement, in the opposite direction

force pulls *downward*, but the sensor indicates a corresponding acceleration *upward*. The output of the accelerometer is then

$$\mathbf{g}^{space} = \begin{pmatrix} 0 \\ 0 \\ -9.81 \end{pmatrix} \text{ms}^{-2} \Rightarrow \mathbf{acc}^{space}_{gravity} = \begin{pmatrix} 0 \\ 0 \\ 9.81 \end{pmatrix} \text{ms}^{-2}. \quad (2.8)$$

IMU Artifacts

The analysis of IMU measurements can be affected by measurement noise, measurement offsets, sensor drifts, external interferences, and by errors in the underlying assumptions. The determination of the exact sensor *position* from IMU signals requires very high-quality sensors. To obtain orientation or position, the measurement signals have to be integrated once or twice, respectively. Thus any bias on the measurement signal effects orientation linearly with time, and position even quadratically with time. For each coordinate direction $i \in \{x, y, z\}$, we have

$$\begin{aligned} vel_i(t') &= \int_{t_0}^{t'} acc_i(t'')dt'' + vel(t_0) \\ pos_i(t) &= \int_{t_0}^{t} vel_i(t')dt' + pos_i(t_0) \\ &= \int_{t_0}^{t}\int_{t_0}^{t'} acc_i(t'')dt''dt' + vel_i(t_0)(t-t_0) + pos_i(t_0). \end{aligned} \quad (2.9)$$

If acceleration has a small bias, i.e., $acc_i(t'')$ is replaced by $acc_i(t'')+\Delta$, Eq. (2.9) leads to

$$pos_i(t) = \int_{t_0}^{t}\int_{t_0}^{t'} acc_i(t'')dt''dt' + vel_i(t_0)(t-t_0) + pos_i(t_0) + \Delta(t-t_0)^2 \,. \quad (2.10)$$

In words, the calculated position $pos_i(t)$ includes a position error of $\Delta * (t - t_0)^2$ which increases quadratically with time.

For the same reason, small errors in the orientation signal induce erroneous position measurements. For example, a tilt error of 0.05° will cause a component of the acceleration due to gravity with a magnitude of 0.0086 ms^{-2} to be projected onto the horizontal axes. This residual bias causes an error in the horizontal position which grows quadratically to 7.7 m after only 30 s!

Unfortunately, signals from the IMUs typically used for human movement recordings often include a significant amount of drift and offset. In order to nevertheless obtain position and orientation values that are close to the actual values, two options are available:

- Additional measurement signals can be used to provide information about the absolute orientation with respect to space. In that case, we are dealing with "over-determined" systems and uncertain information, and we have to find the optimal solution to the given measurement data.
- Or information about external restrictions can be used to improve the accuracy.

The first option is for example used by MEMS devices which also include a magnetometer, which measures the direction of the local magnetic field. This is approximately the direction of the earth magnetic field. This additional information provides a constant orientation and allows to partially compensate erroneous measurement signals. An introduction to the ideas behind "sensory integration" will be presented in Chap. 7.

The second option is to use external information to compensate for measurement errors. For example, one can ask the subject repeatedly to bring the sensor into a well-known position/orientation. Or one can use knowledge about the experiment, e.g., that during walking the movement of the sole of the shoe is typically zero while the shoe is on the ground (Jiménez et al. 2010). Another assumption that is sometimes applied to human movement recordings is that averaged over 10 or more seconds, the mean acceleration points almost exactly downward (Madgwick et al. 2011).

2.2.3 Gyroscopes

Gyroscopes indicate the angular velocity with respect to inertial space, but the values can be expressed in any coordinate system. In most cases, the gyroscope measurements are expressed in a sensor-fixed coordinate system.

MEMS gyroscopes make use of vibrating elements and the "Coriolis effect", which states that in a frame of reference rotating at angular velocity $\boldsymbol{\omega}$, a mass m moving with velocity $\mathbf{v}$ experiences a (fictitious) force $\mathbf{f_c}$, the Coriolis force (see Fig. 2.10):

$$\mathbf{f_c} = 2 * m * (\mathbf{v} \times \boldsymbol{\omega}) \,. \tag{2.11}$$

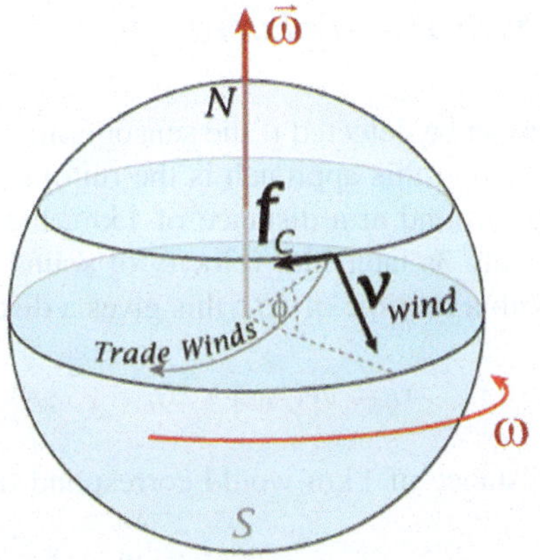

Fig. 2.10 The best-known application example of the Coriolis force are the trade winds. Winds blowing from the poles toward the equator ($\mathbf{v}_{wind}$) change their radius, $r_{earth} * \cos(\phi)$, in an environment rotating with $\boldsymbol{\omega}$ and therefore experience the Coriolis force $\mathbf{f}_C$. This deflects the winds, leading to the commonly observed easterly winds around the equator known as "trade winds"

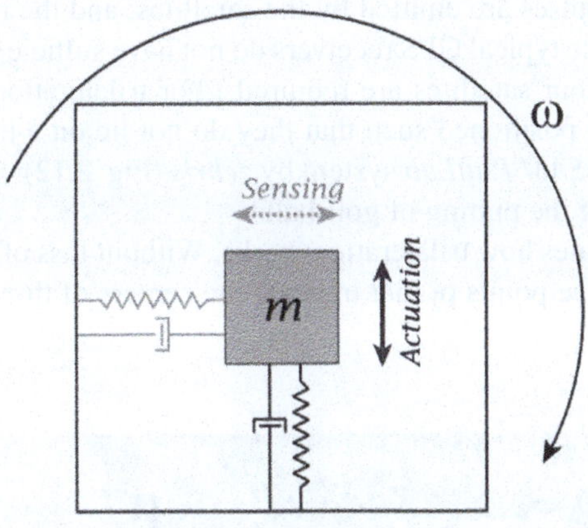

Fig. 2.11 A vibrating mass gyroscope makes use of the Coriolis force: a linear movement along the "Actuation" axis during a rotation causes a torque along the "Sensing" axis

The simplest sensor geometry consists of a single mass which is driven to vibrate along a *drive* or *actuation* axis, as shown in Fig. 2.11. When the gyroscope is rotated, a secondary vibration is induced along the perpendicular *sensing* axis due to the Coriolis force. The angular velocity can be calculated by sensing this secondary vibration.

2.2.4 Ultrasound Sensors—Trilateration

The location of a sensor can be detected if the sensor emits short ultrasound pulses. The physical principle behind this approach is the runtime of sound: for example, if a lightning strikes the ground at a distance of 1 km, the corresponding thunder can be heard approximately 3s later: the velocity of sound in air is approximately $vel_{\text{sound}} = 343\,\text{ms}^{-1}$. With a runtime of Δt_i, this gives a distance r_i of

$$r_i = vel_{\text{sound}} * \Delta t_i \ . \tag{2.12}$$

So a lightning at a distance of 1 km would correspond to a thunder with a time delay of $\Delta t = 292$ ms.

The fact that sound in air propagates isotropically, i.e., in the same way in every direction, can be used to determine the exact location of the sound source in 3-D space if runtime measurements are available from three independent microphones. The method to find the sound source is called "trilateration" (see Fig. 2.13) and requires the measurement of the time delay with three independent detectors. (The same principle is also used for GPS systems, with electromagnetic pulses instead of sound pulses. The pulses are emitted by the satellites, and the runtime is measured by the observer. Since typical GPS receivers do not have sufficiently accurate clocks, in practise at least four satellites are required.) For trilateration to work, the three detectors have to be positioned such that they do not lie on a line. An example for such a system is the *SAM PuttLab* system by *zebris* (Fig. 2.12), which is a dedicated system for analyzing the putting of golf balls.

Figure 2.13 indicates how trilateration works. Without loss of generality, it can be assumed that the three points $\mathbf{p_i}$ that indicate the centers of three spheres with radii r_i, all lie in a plane.

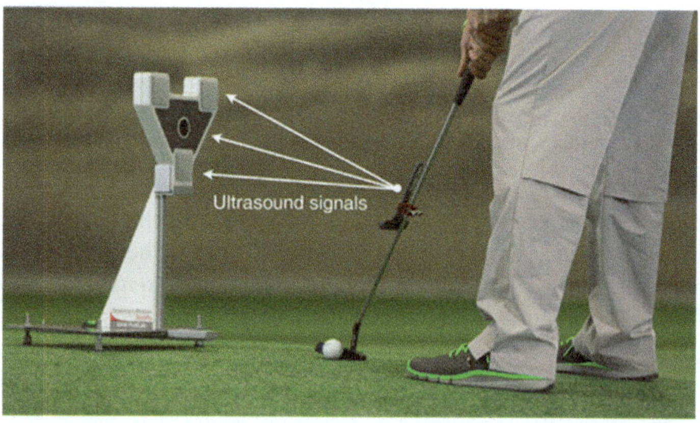

Fig. 2.12 The SAM PuttLab by *Science & Motion Sports* (With kind permission from http://www.scienceandmotion.com)

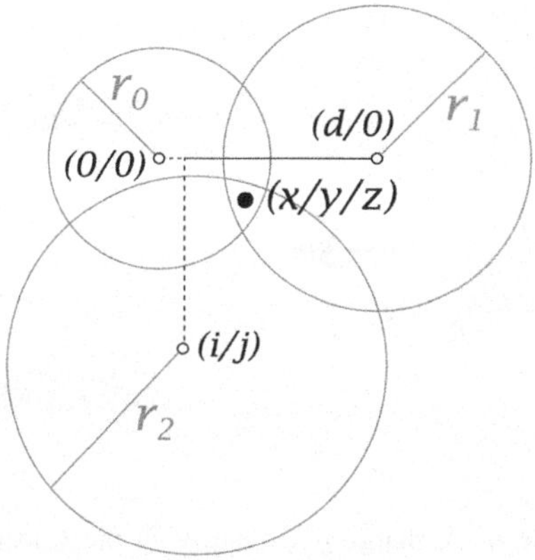

Fig. 2.13 The plane $z = 0$, showing from the upper left the three sphere centers $\mathbf{p_0}, \mathbf{p_1}, \mathbf{p_2}$ (open dots, for clarity no labels); their x, y-coordinates; and the corresponding sphere radii, r_0, r_1, and r_2. The two intersections of the three sphere surfaces are directly in front and directly behind the point designated (x, y, z) in the z = 0 plane

- $\mathbf{p_0} = (0/0/0)$ forms the center of the coordinate system,
- the line from $\mathbf{p_0}$ to $\mathbf{p_1} = (d/0/0)$ forms the positive x-axis, and
- the plane spanned by $\mathbf{p_0}, \mathbf{p_1}, \mathbf{p_2} = (i/j/0)$ is the x–y-plane.

The radius r_i of each sphere can be determined from the propagation time of the signal. For example, with an ultrasound-based system such as the one shown in Fig. 2.12, a propagation time of $\Delta t = 2$ ms would thus give a radius of approximately 69 cm.

It can then be shown that the intersection of the three spheres has the coordinates

$$
\begin{aligned}
x &= \frac{r_0^2 - r_1^2 + d^2}{2d} \\
y &= \frac{r_0^2 - r_2^2 - x^2 + (x-i)^2 + j^2}{2j} = \frac{r_0^2 - r_2^2 + i^2 + j^2}{2j} - \frac{i}{j}x \\
z &= \pm\sqrt{r_0^2 - x^2 - y^2}.
\end{aligned}
\tag{2.13}
$$

2.2.5 Magnetic Field Sensors

Static Magnetic Fields

The IMU inertial sensors (accelerometers and gyroscopes) only provide information about the *change* of position and orientation. But due to the earth's magnetic field, we have at least a small magnetic field at every location on the planet, so information about the *absolute* current orientation of the sensor with respect to space can be obtained by measuring the orientation of that field.

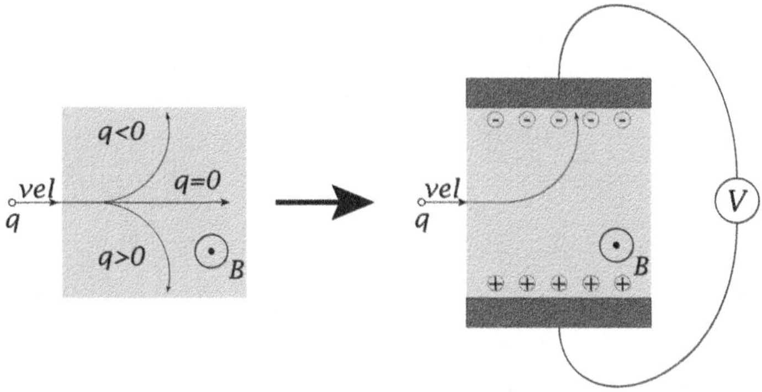

Fig. 2.14 Lorentz force (left): A charge q moving with the velocity vel in a magnetic field B perpendicular to the velocity vel experiences the Lorentz force. This force depends on the charge q, and the particle gets correspondingly deflected. **Hall effect (right)**: The deflected charges accumulate on electrodes, resulting in a voltage V that can be measured with a volt meter

Static magnetic fields can be measured with the *Hall effect*, which can produce a voltage difference (the *Hall voltage*) across an electrical conductor (Fig. 2.14, right).

The Hall effect is based on the *Lorentz force*, the combination of the electric and magnetic forces on a moving point charge due to electromagnetic fields. With an electrical field **E** and a magnetic field **B**,[3] the Lorentz force is

$$\mathbf{f} = q * \mathbf{E} + q * \mathbf{v} \times \mathbf{B}, \tag{2.14}$$

which reduces to

$$\mathbf{f} = q * \mathbf{v} \times \mathbf{B}, \tag{2.15}$$

in the absence of a static electric field (Fig. 2.14, left). With a magnetic field **B** perpendicular to the electric current **v**, this force is transverse to the current in the conductor, producing the Hall effect as indicated in Fig. 2.14 (right).

Application: Earth Magnetic Field Typically, the orientation of the local magnetic field is determined by the earth magnetic field. Figure 2.15 shows a sketch of the *inclination* of the earth magnetic field, i.e., the angle between the magnetic field lines and the local earth-horizontal plane. In Austria, for example, the inclination of the earth magnetic field is approximately 64° (i.e., down), similar to the inclination in San Francisco or New York. The *declination* of the earth magnetic field is the deviation of the field relative to true north ("Ng" in Fig. 2.15), and is positive for an eastward deviation from true north. It is about 3° in Austria (i.e., slightly east relative to true north), 13° in San Francisco, and −10° in New York. However, in the presence of devices that produce magnetic fields, such as old "cathode ray tube"

[3]Note that **E** and **B** are vectors, not matrices. Since they are conventionally capitalized, they are also capitalized here.

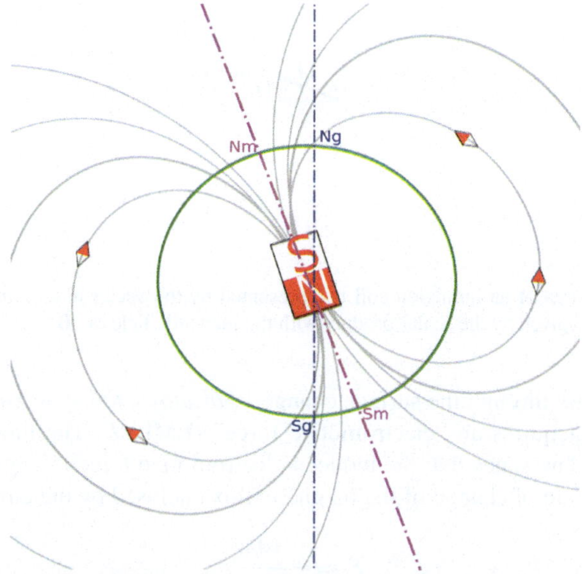

Fig. 2.15 Sketch of the variation between magnetic north (Nm) and "true" north (Ng) (from Wikipedia, "Geomagnetisme.png")

(CRT) screens, the local magnetic field may be strongly distorted. If you use the magnetic field data in your analysis, double check the local magnetic field signals!

Dynamic Magnetic Fields

To measure the orientation of a sensor in an oscillating magnetic field the "induction effect" can be used.

The *magnetic flux* through the coil, which is sometimes referred to as "search coil", is defined by a surface integral:

$$\Phi_B = \int_{\Sigma} \mathbf{B} \cdot d\mathbf{a}, \tag{2.16}$$

where $d\mathbf{a}$ is an element of the surface Σ enclosed by the wire loop, and $\mathbf{B}$ is the magnetic field. The dot product $\mathbf{B} \cdot d\mathbf{a}$ corresponds to the magnetic flux through the surface element $d\mathbf{a}$. In more visual terms, the magnetic flux through the wire loop is proportional to the number of magnetic flux lines that pass through the loop (Fig. 2.16).

If the magnetic field is approximately uniform within the loop, Eq. (2.16) reduces to

$$\Phi_B = \mathbf{B} \cdot \mathbf{a} = |\mathbf{B}| * |\mathbf{a}| * \cos(\alpha) \tag{2.17}$$

where $\mathbf{a}$ is a vector perpendicular to the loop, with a length given by the area enclosed by the loop.

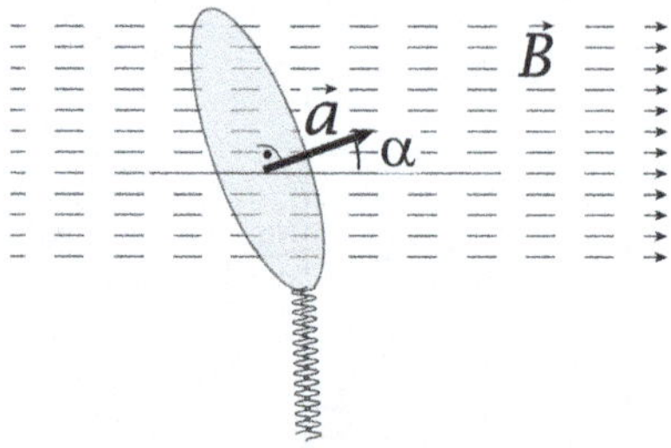

Fig. 2.16 If the area of an induction coil is represented by the vector **a**, then the magnetic flux through the coil is given by the scalar product with the magnetic field, **a** · **B**

When the flux through the surface changes, *Faraday's law of induction* says that the wire loop acquires an "electromotive force" (EMF) $\mathcal{E}$. The most widespread version of this law states that the induced electromotive force in any closed circuit is equal to the rate of change of the magnetic flux enclosed by the circuit:

$$\mathcal{E} = -\frac{d\Phi_B}{dt}. \tag{2.18}$$

This effect is exploited by different measurement systems. In clinical applications, the gold standard for the measurement of 3-D eye orientations was for a long time the "scleral search coil" (Fig. 2.17, left). In these systems, large magnetic field frames (Fig. 2.17, right) generate an approximately homogeneous oscillating magnetic field at the center of the coil frame. As explained in Sect. 3.6.3, this allows the recording of 3-D eye movements with high spatial and temporal precision (Collewijn et al. 1985) (Fig. 2.18). Nowadays, video-based systems have reached a comparable accuracy and resolution (van der Geest and Frens 2002), and have become the dominant measurement devices for eye movements recordings.

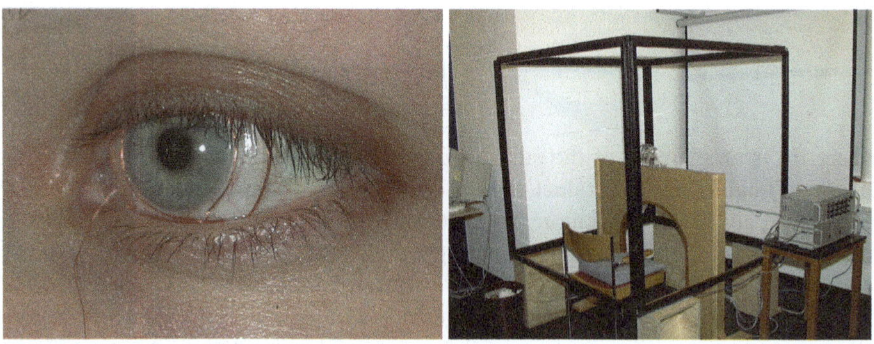

Fig. 2.17 Left: Scleral search coil from the company *Skalar*. Right: External magnetic field frame

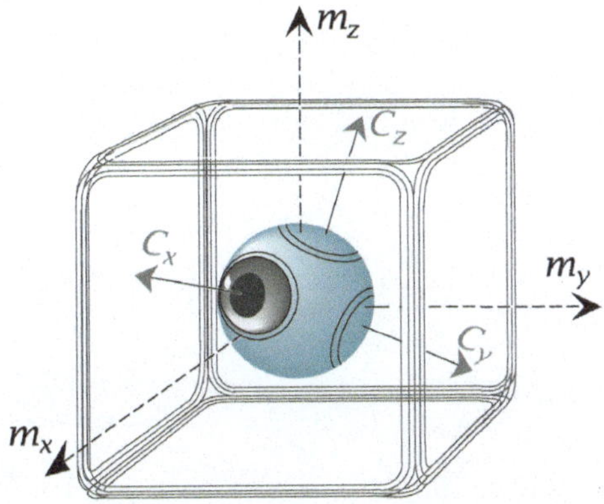

Fig. 2.18 An idealized experimental setup with three orthogonal magnetic fields and three orthogonally mounted induction coils. The induction coils are rigidly attached to the eye, and the coil vectors $[\mathbf{c_x}\ \mathbf{c_y}\ \mathbf{c_z}]$ are parallel to the axes of the eye-fixed coordinate system $[\mathbf{b_x}\ \mathbf{b_y}\ \mathbf{b_z}]$. The magnetic fields $[\mathbf{m_x}\ \mathbf{m_y}\ \mathbf{m_z}]$ are parallel to the space-fixed coordinate system $[\mathbf{s_x}\ \mathbf{s_y}\ \mathbf{s_z}]$. See also the application example for the analysis of signals from search coils in Sect. 3.6.3

Fig. 2.19 Induction systems with inhomogeneous magnetic fields generated by a small magnetic field generator can determine not only the orientation of the sensor in space but also its position. Left: The *Aurora* system from NDI. Right: the corresponding, tiny sensor coils (inside the tubular black shielding cover)

Other movement recording systems build on a compact generator of an inhomogeneous magnetic field (Fig. 2.19, left). Smart algorithms make use of the field inhomogeneity to determine position *and* orientation of sensor coils (Kindratenko 2000). Thereby, the size of the search coils is small enough that they can be used even for endoscopy applications (Fig. 2.19, right).

Chapter 3
Rotation Matrices

3.1 Introduction

Six parameters (degrees of freedom) are required and sufficient to completely describe the movement of an object in space: three describe the 3-D position of the object, and three the 3-D orientation, often referred to as "attitude" in aeronautics. When describing movements that are less than a few kilometers, we often use space-fixed, Cartesian coordinate systems. In these systems, the orientation of each axis is the same for each point in space, and for all time. For typical industrial applications and for movement measurements, a system fixed with respect to the surface of the earth can be regarded as space-fixed, as a convenient substitute of a true "inertial system".[1] These coordinate systems are sometimes called local-level-local-North, assuming a flat earth. For larger distances, when the curvature of the earth becomes significant, the direction of "up" starts to depend on the location, and Cartesian coordinate systems are no longer useful. An example of such a non-cartesian coordinate system is the geodetic latitude/longitude/height coordinate system.

After choosing an arbitrary point in space as the *reference position*, the *position* of each point is defined by three translations away from the coordinate center, e.g., forward, left, and up. Here, it is worth pointing out a seemingly obvious fact: the final location of the object is independent of the sequence of these translations. If we move first 10 m right and then 15 m forward, we end up in the same location as if we had moved first 15 m forward, and then 10 m right. This property is referred to as the *commutativity* of translations (see also Fig. 5.1).

The description of orientation is done in a similar way (Fig. 3.1). First, an arbitrarily chosen orientation is defined as *reference orientation*. Once that is done, any other orientation can be described by three parameters: an object can not only be

[1] An inertial frame is a frame of reference in which a body remains at rest or moves with a constant linear velocity unless acted upon by forces. An inertial reference frame does not have a single, universal coordinate system attached to it: positional values in an inertial frame can be expressed in any convenient coordinate system. In other words, an inertial frame is a frame of reference where the laws of inertia apply—there is no requirement for specific coordinates.

T. Haslwanter, *3D Kinematics*, https://doi.org/10.1007/978-3-319-75277-8_3

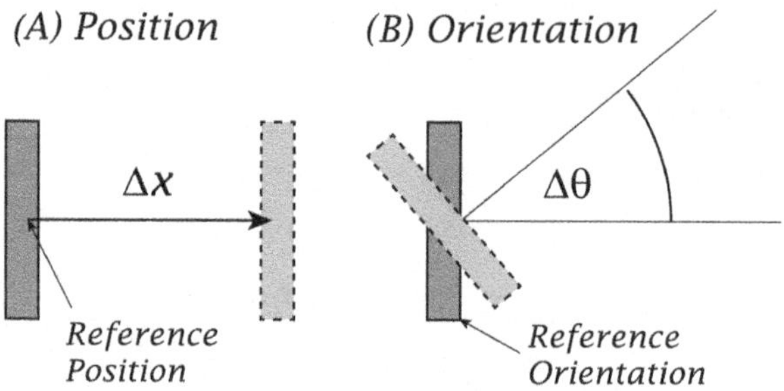

Fig. 3.1 Similarity between one-dimensional translations (left) and single-axis rotations (right): Both require the selection of a reference, and both are characterized by a single parameter

translated along each of the three coordinate axes, but it can also be rotated about each of these axes. To describe the three-dimensional orientation, two inherently different approaches can be taken. The first approach is based on *Euler's Theorem*, which states that for every two orientations of an object, the object can always move from one to the other by a single rotation about a fixed axis (Euler 1775). In that case, the axis of the rotation is defined by two parameters, and the magnitude of the rotation defines the third parameter. The second approach is to describe the rotation from the reference orientation to the current orientation through three consecutive rotations about well-defined, hierarchically nested coordinate axes (e.g., Goldstein 1980). For a long time, this has been the most common approach to characterize orientation in three dimensions. A detailed analysis of all the rotation angle sequences is given by (Diebel 2006).

The following section will deal with this three-rotation description of 3-D orientation, while the approach based on Euler's Theorem will be explained in detail in Chap. 4 ("Quaternions").

3.2 Rotations in a Plane

A simple rotation of a point **p** in a plane can be uniquely described in two ways:

- In *Cartesian coordinates* through a rotation matrix.
- In *polar coordinates*, through an angle θ characterizing the rotation about an axis perpendicular to the plane.

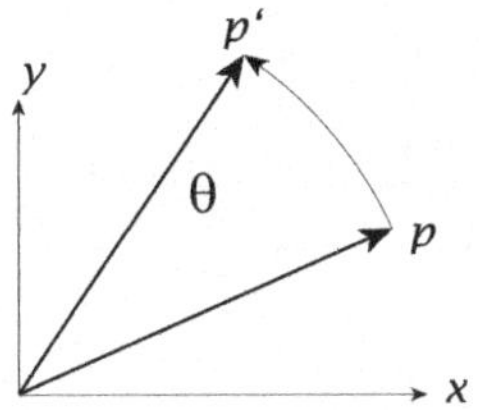

Fig. 3.2 Rotation in a plane

3.2.1 Rotation in Cartesian Coordinates

In two-dimensional Cartesian coordinates, a point $\mathbf{p}$ is defined by its two coordinate components $\mathbf{p} = \begin{pmatrix} x \\ y \end{pmatrix}$. When that point is rotated by an angle θ into a new point $\mathbf{p}' = \begin{pmatrix} x' \\ y' \end{pmatrix}$ (Fig. 3.2), the coordinates of $\mathbf{p}'$ are given by[2]

$$\begin{pmatrix} x' \\ y' \end{pmatrix} = \begin{bmatrix} \cos\theta & -\sin\theta \\ \sin\theta & \cos\theta \end{bmatrix} \cdot \begin{pmatrix} x \\ y \end{pmatrix} . \tag{3.1}$$

Defining the "rotation matrix" $\mathbf{R}$ as

$$\mathbf{R} = \begin{bmatrix} \cos\theta & -\sin\theta \\ \sin\theta & \cos\theta \end{bmatrix} , \tag{3.2}$$

Equation (3.1) can be rewritten as

$$\mathbf{p}' = \mathbf{R} \cdot \mathbf{p}. \tag{3.3}$$

Note that the columns of the rotation matrix are equivalent to the basis vectors of the space-fixed coordinate system ($\mathbf{n_x}$, $\mathbf{n_y}$) rotated by the angle θ (Fig. 3.3)! Or in other words, the rotation matrix is the projection of the rotated unit vectors onto the coordinate axes. The rotation matrix is therefore sometimes also referred to as the "direction cosine matrix (DCM)".

$$\mathbf{R} = \mathbf{R} \cdot \left[\mathbf{n_x}\ \mathbf{n_y}\right] = \left[\mathbf{n_x'}\ \mathbf{n_y'}\right]. \tag{3.4}$$

[2]Note for Matlab users: here and in the following, the dash in $\mathbf{p}'$ does NOT mean the vector $\mathbf{p}$ transposed, but rather the vector $\mathbf{p}$ rotated!

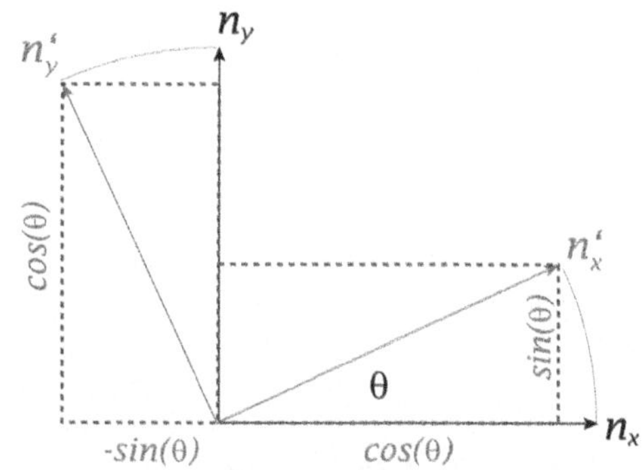

Fig. 3.3 Rotation matrix: projection in 2-D

3.2.2 Rotation in Polar Coordinates

Every complex number has a real and an imaginary part

$$c = x + j * y$$

and therefore can be represented by a vector in the (x, y)-plane (Fig. 3.4). And since

$$e^{j\theta} = \cos(\theta) + j * \sin(\theta), \tag{3.5}$$

every complex number can also be represented by a magnitude r and an angle θ:

$$c = r * e^{j\theta} = r * (\cos(\theta) + j * \sin(\theta)), \tag{3.6}$$

where

$$r = \sqrt{Re^2 + Im^2} \tag{3.7}$$

$$\theta = \arctan\left(\frac{Im}{Re}\right). \tag{3.8}$$

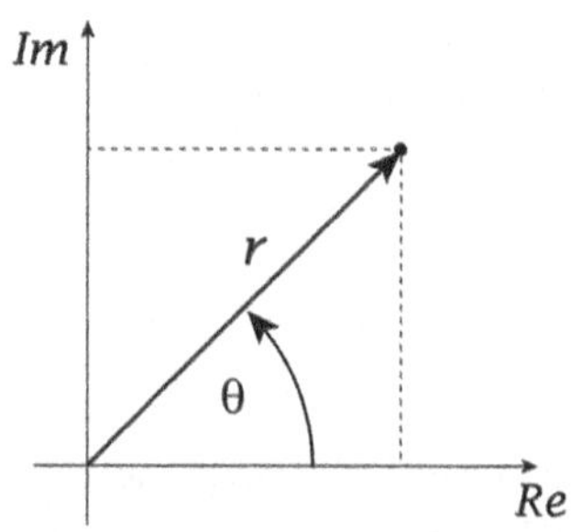

Fig. 3.4 Complex number, in polar coordinates (r, θ)

As a result, a rotation of a 2-D-vector, expressed as a complex number c, by an angle ϕ, can be written as

$$c' = e^{j\phi} * c = e^{j\phi} * (r * e^{j\theta}) = r * e^{j(\phi+\theta)}. \tag{3.9}$$

Note: In mathematics and physics, the square root of -1 is typically denoted with i, whereas in many technical areas j is used. In both Python and Matlab, j can be used:

```
x = -1+0j
np.sqrt(x)
>>> 1j
```

In polar coordinates, the similarity between one-dimensional translations and single-axis rotations becomes obvious (Fig. 3.1).

3.2.3 *Application: Orienting an Object in a Plane*

Task: If a gun originally pointing straight ahead along the $+x$ axis is to shoot at a target at $\mathbf{P} = (x, y)$, by which amount does the gun have to rotate to point at that target (Fig. 3.5)?

Solution:

The gun barrel originally points straight ahead, so the direction of the bullet aligns with $\mathbf{n_x}$.

The rotation of the gun is described by the rotation matrix $\mathbf{R} = \begin{bmatrix} \cos\theta & -\sin\theta \\ \sin\theta & \cos\theta \end{bmatrix} = \begin{bmatrix} \mathbf{n'_x} & \mathbf{n'_y} \end{bmatrix}$. The direction of the gun barrel after the rotation is given by $\mathbf{n'_x} = \frac{\mathbf{p}}{|\mathbf{p}|}$, which is also the first column of the rotation matrix $\mathbf{R}$.

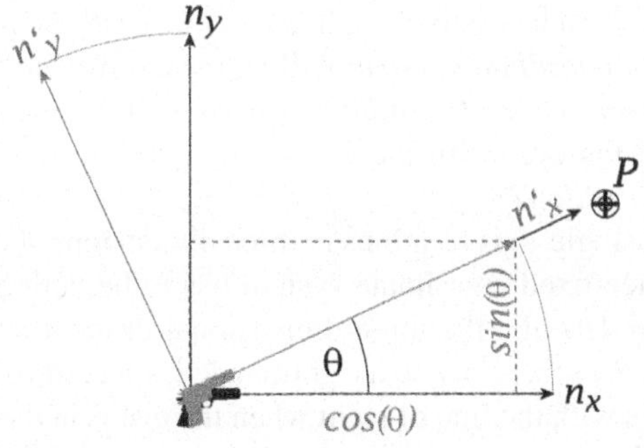

Fig. 3.5 Targeting with one degree-of-freedom (DOF)

Combining these two equations leads to the vector equation

$$\begin{pmatrix} \cos\theta \\ \sin\theta \end{pmatrix} = \frac{1}{\sqrt{x^2+y^2}} \begin{pmatrix} x \\ y \end{pmatrix}.$$

The y-component of this vector equation is $\sin\theta = \frac{y}{\sqrt{x^2+y^2}} \rightarrow \theta = \arcsin\frac{y}{\sqrt{x^2+y^2}}$

Note: For small angles ($\theta \ll 1$), $\sin(\theta)$ and $\cos(\theta)$ can be expanded with a Taylor series, and one obtains in a linear approximation

$$\begin{aligned} \sin(\theta) &\approx \theta + 0(\theta^2), \text{ and} \\ \cos(\theta) &\approx 1 + 0(\theta^2). \end{aligned} \tag{3.10}$$

As a result, for small angles, numerical errors are minimized by calculating the angle from measurements related to $\sin(\theta)$ as shown above. For angles around 90° the component proportional to $\cos(\theta)$ should be used, to minimize numerical errors in the computation of θ.

3.3 Rotations About Coordinate Axes in 3-D

3.3.1 3-D Rotations About Coordinate Axes

The same ideas as described above for rotations in a plane can be applied to three dimensions, leading to the 3-D rotation matrix

$$\mathbf{R} = \left[\mathbf{n}'_x\ \mathbf{n}'_y\ \mathbf{n}'_z\right], \tag{3.11}$$

where the $\mathbf{n}'_i$ are column vectors. In the following, eye movements will often be used as an example, as they can be easily visualized. In that case, the *reference coordinate system* or *space-fixed coordinate system* will be the coordinate system provided by the head, and the *eye(body)-fixed coordinate system* will here be a coordinate system fixed to the body of the eye, with the x-axis aligned with the line of sight (often referred to as *gaze direction*).

In order to define single-axis rotations in three dimensions about coordinate axes, first an external, space-fixed coordinate system has to be defined and a body-fixed coordinate system to describe the three-dimensional orientation of the object with respect to space. Let $\mathbf{S} = \left[\mathbf{s_x}\ \mathbf{s_y}\ \mathbf{s_z}\right]$ be a right-handed, space-fixed coordinate system such that $\mathbf{s_x}$ coincides with the line of sight when the eye is in the reference position, $\mathbf{s_y}$ with the interaural axis (i.e., left-right), and $\mathbf{s_z}$ with earth vertical (Fig. 3.6a).

Let $\mathbf{B} = \left[\mathbf{b_x}\ \mathbf{b_y}\ \mathbf{b_z}\right]$ (note: $\mathbf{b}_i$ $i = x, y, z$ are column vectors!) denote a right-handed body-fixed coordinate system (i.e., it moves with the object, here the eye) such that $\mathbf{B}$ coincides with the space-fixed coordinate system $\mathbf{S}$ when the eye/body

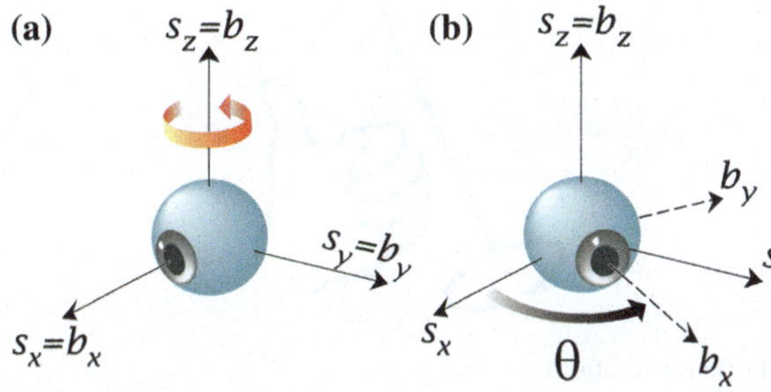

Fig. 3.6 Horizontal rotation of the eye about the space-fixed axis $\mathbf{s_z}$ by an angle θ from the reference orientation (**a**) to a new orientation (**b**)

is in the reference orientation. Any horizontal rotation of the body-fixed coordinate system (and thus of the object) from the reference orientation to a new orientation, as indicated in Fig. 3.6b, can be described by

$$\mathbf{b_i} = \mathbf{R} \cdot \mathbf{s_i}, \mathrm{i} = x, y, z\,, \tag{3.12}$$

or, equivalently

$$\mathbf{B} = \mathbf{R} \cdot \mathbf{S}. \tag{3.13}$$

The rotation matrix **R** describes the orientation of the eye/body (**B**) relative to the head (**S**).

For a rotation of a point about a vertical axis in a space-fixed coordinate system, as indicated in Fig. 3.6b, the matrix **R** describes a rotation about $\mathbf{s_z}$ by an angle of θ. This matrix, which we here call $\mathbf{R}_z(\theta)$, is given by

$$\mathbf{R}_z(\theta) = \begin{bmatrix} \cos\theta & -\sin\theta & 0 \\ \sin\theta & \cos\theta & 0 \\ 0 & 0 & 1 \end{bmatrix}. \tag{3.14}$$

In the same way, vertical rotation of a point about $\mathbf{s_y}$ in a space-fixed coordinate system by an angle of ϕ can be described by

$$\mathbf{R}_y(\phi) = \begin{bmatrix} \cos\phi & 0 & \sin\phi \\ 0 & 1 & 0 \\ -\sin\phi & 0 & \cos\phi \end{bmatrix}, \tag{3.15}$$

and torsional rotation of a point about $\mathbf{s_x}$ in a space-fixed coordinate system by an angle of ψ by

$$\mathbf{R}_x(\psi) = \begin{bmatrix} 1 & 0 & 0 \\ 0 & \cos\psi & -\sin\psi \\ 0 & \sin\psi & \cos\psi \end{bmatrix}. \tag{3.16}$$

Fig. 3.7 Right-hand rule for rotations

With these definitions, and with the positive (x, y, z)-axis pointing forward/left-/up, respectively, positive θ, ϕ, and ψ values correspond to leftward, downward, and clockwise movements. Depending on the context, these three rotations are either called *yaw, pitch* and *roll* angles (in nautical applications, or in eye movement research), or—equivalently—*heading, elevation* and *banking* angles (in aerospace engineering). In the latter case, the term *attitude* is used to characterize 3-D orientation.

Notes:

- The direction of positive rotations can easily be remembered with the right-hand rule (Fig. 3.7): if a body is gripped with the right hand and rotated in the direction of the curled fingers, the direction of the thumb determines the sign of the rotation. With the coordinate system as defined in Fig. 3.6, rotations to the left, downward(!), and clockwise (as seen from the user) are positive.
- Care has to be taken with the implementation of rotations, such as in Eq. (3.13), if the data are in row format. For example, if data are stored as

$$\mathbf{Data} = \begin{bmatrix} x_0 & y_0 & z_0 \\ x_1 & y_1 & z_1 \\ x_2 & y_2 & z_2 \\ x_3 & y_3 & z_3 \\ . & . & . \\ . & . & . \\ . & . & . \end{bmatrix} \quad (3.17)$$

then using the matrix notation, the rotation of these data has to be implemented as

$$\mathbf{Data}' = \left[\mathbf{R} \cdot \mathbf{Data}^T\right]^T = \mathbf{Data} \cdot \mathbf{R}^T , \quad (3.18)$$

because

$$[\mathbf{A} \cdot \mathbf{B}]^T = \mathbf{B}^T \cdot \mathbf{A}^T . \quad (3.19)$$

3.3.2 *Rotations of Objects Versus Rotations of Coordinate Systems*

The next step is conceptually trivial, but its ramifications have caused consternation among generations of scientists.

Figure 3.8a shows the reference setup, where a picture is taken of a cat, with the cat's nose in the center of the image. In Fig. 3.8b, *the cat* has been rotated by 25°, and its nose is now at the lower edge of the captured image. In Fig. 3.8c, the cat remains stationary, but now *the camera* coordinate system is rotated by 25°, this time in the opposite direction. The image of the cat looks exactly the same as in Fig. 3.8b.

From the definition of our coordinate system (forward, left, and up are the positive directions), and the choice of the right-hand rule, a "downward" rotation has to be positive. And in both cases (Fig. 3.8b, c), the final relative orientation between camera and cat is the same. But now the catch is: should the relative movement shown in Fig. 3.8b and c be labeled a *downward* rotation, because the cat is rotated downward relative to the camera? Or should it be labeled an *upward* rotation, because the camera is rotated up with respect to the surroundings? In situations where the measurement setup is stationary, such as in Fig. 3.8b, it makes sense to call this relative rotation a *downward* rotation. (This is the convention used in this book, where **R** describes the rotation of an object relative to a space-fixed coordinate system.)! But when a constant environment is observed from a moving object, as in Fig. 3.8c, it makes more sense to call this relative rotation an *upward* rotation, because the camera rotates *upward.* (That is the convention often used in theoretical physics, theoretical mechanics, and aeronautics, where a fixed world and fixed events are observed from different, moving reference systems.) It would be appropriate to describe, for example, the orientation of a moving aeroplane relative to a fixed-world environment.)

There exists no *correct* choice here, only two different options. Which one is chosen depends on the field of research and the application.

When comparing the equations from this book with other literature, carefully check if the definition of the rotation matrices is consistent with Eqs. (3.14)–(3.16). If this is the case, everything is fine. But if the definition of the rotation matrices

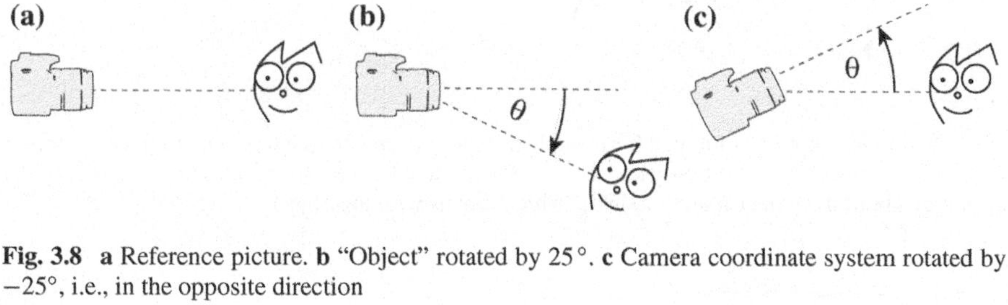

Fig. 3.8 **a** Reference picture. **b** "Object" rotated by 25 °. **c** Camera coordinate system rotated by −25°, i.e., in the opposite direction

is the transposed of Eqs. (3.14)–(3.16), the alternate choice has been made, which affects all subsequent equations![3]

Important Note
Even though I am repeating myself: Care has to be taken, because the exact form of the rotation matrices depends on the definition of **R**. Technical applications often use rotations of the coordinate system for the definition of the rotation matrix, and the signs of the angles are inverted compared to our definitions in Eqs. (3.14)–(3.16). In those applications, for example in the excellent summary by (Diebel 2006), the rotation matrices are exactly the transposed versions of the matrices used here!!

3.4 Combined Rotations

For rotations about a single axis, no distinction has to be made between rotations about body-fixed or space-fixed axes. Since the body-fixed and space-fixed coordinate systems coincide when the object is in the reference position, the axis about which the object rotates is the same in the body-fixed and space-fixed system. But this is no longer the case for combined rotations about different axes. For such rotations the elements of **R** are no longer determined by the relatively simple formulas in Eqs. (3.14)–(3.16).

The example in Fig. 3.9 may help to better understand the problem: how should we distinguish between a downward movement of the object by a rotation about the space-fixed axis $\mathbf{s_y}$ (as shown in Fig. 3.9a) and a downward movement by a rotation about the rotated, body-fixed axis $\mathbf{b_y}$ (Fig. 3.9b)?

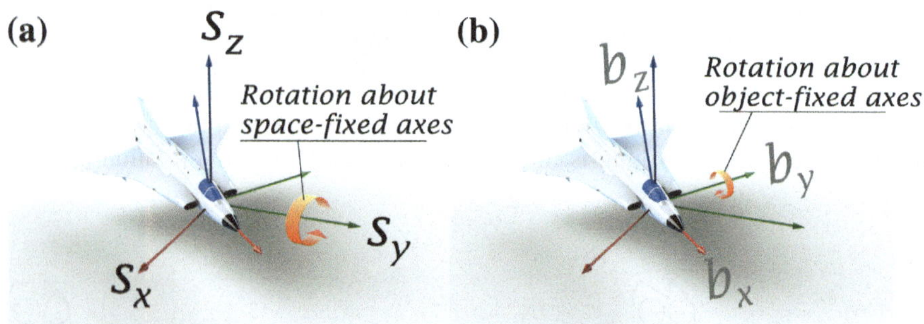

Fig. 3.9 In describing a combined horizontal–vertical movement, one has to distinguish clearly if the vertical movement is (**a**) a rotation about the space-fixed y-axis $\mathbf{s_y}$, which remains fixed, or (**b**) a rotation about the object-fixed y-axis $\mathbf{b_y}$, which moves with the object

[3]Appendix A.3.3 contains the proof that the body-fixed representation of rotations uses the inverse (i.e., the transpose) rotation matrix compared to the space-fixed representation.

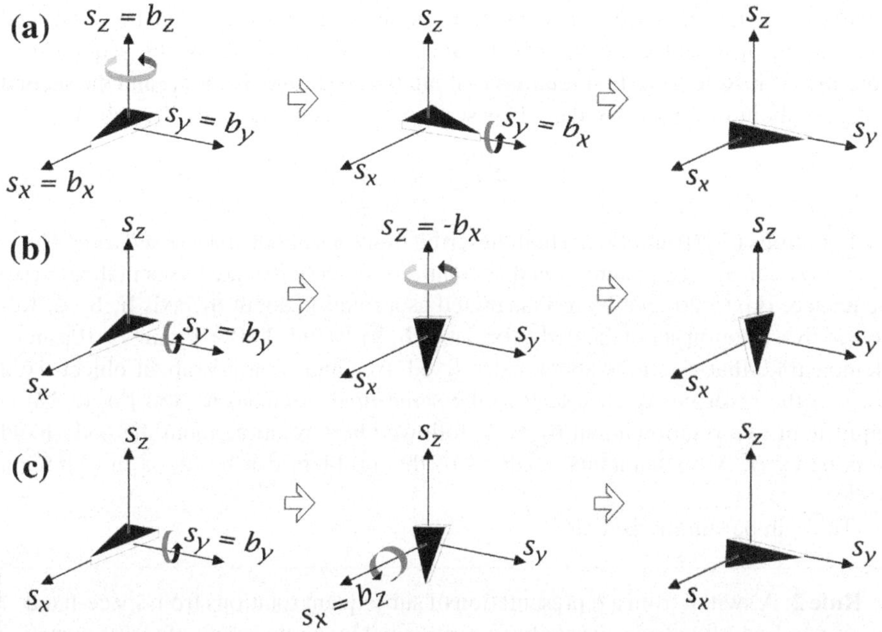

Fig. 3.10 Sequences of two rotations. **(a) Space-fixed, $\mathbf{s_z}$ :: $\mathbf{s_y}$** : 90° rotation about the vertical axis $\mathbf{s_z}$, followed by a 90° rotation about the horizontal axis $\mathbf{s_y}$. **(b) Space-fixed, $\mathbf{s_y}$ :: $\mathbf{s_z}$** : 90° rotation about the horizontal axis $\mathbf{s_y}$, followed by a 90° rotation about the vertical axis $\mathbf{s_z}$. **(c) Body-fixed, $\mathbf{b_y}$ :: $\mathbf{b_z}$** : 90° rotation about the body-fixed axis $\mathbf{b_y}$, followed by a 90° rotation about the body-fixed axis $\mathbf{b_z}$. The final orientation is the same as in **(a)**. Body-fixed axes and space-fixed axes are superposed because the size of the rotations in this example is exactly 90°

Mathematically, the difference between rotations in space-fixed coordinates and body-fixed coordinates lies in the sequence in which the rotations are executed. This is illustrated in Fig. 3.10. The upper column (Fig. 3.10a) shows a rotation of an object about $\mathbf{s_z}$ by $\theta = 90°$, followed by a rotation about the space-fixed axis $\mathbf{s_y}$ by $\phi = 90°$. Mathematically, this is described by

$$\mathbf{b_i} = \mathbf{R}_y(\phi) \cdot \mathbf{R}_z(\theta) \cdot \mathbf{s_i} \tag{3.20}$$

with $\theta = \phi = 90°$.

Note: The rotation that is executed first is on the *right-hand side*, because this is the first matrix to act on the object to be rotated:

$$\mathbf{R}_y(\phi) \cdot (\mathbf{R}_z(\theta) \cdot \mathbf{s_i}) = (\mathbf{R}_y(\phi) \cdot \mathbf{R}_z(\theta)) \cdot \mathbf{s_i} \, . \tag{3.21}$$

This leads to

> **Rule 1**: Subsequent rotations are written *right-to-left*.

Inverting the sequence of two rotations about space-fixed axes changes the final orientation of the object. This can be seen in Fig. 3.10b, where the sequence of rotations is inverted: the first rotation is about the space-fixed axis $\mathbf{s_y}$, and the second rotation about the space-fixed $\mathbf{s_z}$. This sequence is mathematically described by

$$\mathbf{b_i} = \mathbf{R}_z(\theta) \cdot \mathbf{R}_y(\phi) \cdot \mathbf{s_i}. \tag{3.22}$$

Equations (3.20) and (3.22) both describe rotations about space-fixed axes. However, they can also be re-interpreted as rotations about body-fixed axes in the reverse sequence: Eq. (3.20) can be re-interpreted as a rotation about the axis $\mathbf{b_y}$ by ϕ, followed by a rotation about the body-fixed axis $\mathbf{b_z}$ by θ (Fig. 3.10c). Figures 3.10a and c demonstrate that rotations about space-fixed axes and rotations about object-fixed axes in the reverse sequence lead to the same final orientation. And Eq. (3.22) is equivalent to a rotation about $\mathbf{b_z}$ by θ, followed by a rotation about the body-fixed axis $\mathbf{b_y}$ by ϕ. A mathematical analysis of this problem can be found in (Altmann 1986).

This can be summarized as

> **Rule 2**: A switch from a representation of subsequent rotations from space-fixed axes to body-fixed axes has to be accompanied by an inversion of the sequence of the rotation matrices.

This also gives the answer to the problem raised by Fig. 3.9: the combination of two rotations about the space-fixed axes $\mathbf{s_z}$ and $\mathbf{s_y}$, as shown in Fig. 3.9a, is mathematically described by Eq. (3.20), while the combination of two rotations about the object-fixed axes $\mathbf{b_z}$ and $\mathbf{b_y}$, as shown in Fig. 3.9b, is described by Eq. (3.22).

Rotations about space-fixed axes are often called "rotations of the object" or "active rotations", since in successive rotations only the object is rotated, and the axes of the successive rotations are unaffected by the preceding rotations of the object. Rotations about object-fixed axes are often referred to as "rotations of the coordinate system" or "passive rotations", since each rotation changes the coordinate axes about which the next rotations will be performed.

3.4.1 3-D Orientation with Sequential Rotations

Systems that use such a combination of three rotations for the description of the orientation can be demonstrated with *gimbal systems*. A gimbal is a ring or a frame that is suspended so it can rotate about an axis. Gimbals are typically nested one within another to accommodate rotation about multiple axes, and the hierarchy of sequential rotations is automatically implemented. As pointed out in the Chap. 1, we will by default use an inertial coordinate system where the positive x-, y-, and z-axes point forward, left, and up.

3.4.1.1 Nautical Sequence

In aeronautics (Kuipers 1999) and maritime applications, the yaw–pitch–roll sequence of rotations, which we will refer to as "nautical sequence", is very common.

The **first** rotation is about the (vertical) z-axis. This movement is called a "yaw movement", and the corresponding angle is sometimes referred to as "heading angle". According to the right-hand rule, a positive rotation rotates the nose of the airplane to the left.

The **second** movement is about the once-rotated, body-fixed y-axis (when the object is an airplane, about the line connecting the wings). The movement is called a "pitch movement", and the corresponding angle is referred to as "elevation angle".

The **last** rotation is about the twice-rotated x-axis (when the object is an airplane, about the longitudinal axis). This movement is called "roll movement", and the corresponding angle is the "banking angle". If the rotated object is a camera or an eye, this roll rotation will not change the line of sight or gaze direction, but it will rotate the image.

This sequence has first been used by the German doctor and physiologist Adolf Fick (Fick 1854), who worked on eye movements and who also invented the first contact lenses worn by patients. In eye movement research, the yaw, pitch, and roll angles for this sequence are therefore often referred to as "Fick angles". The yaw and pitch angles together determine the line of sight, and the corresponding direction is called the "gaze direction".

The left illustration in Fig. 3.11 shows a gimbal which corresponds to the nautical sequence of rotations. The angles of the nautical sequence will be denoted by the subscript "N" (θ_N, ϕ_N, ψ_N). The rotation matrix corresponding to the nautical sequence of rotations is

$$\mathbf{R}_{nautical} = \mathbf{R}_z(\theta_N) \cdot \mathbf{R}_y(\phi_N) \cdot \mathbf{R}_x(\psi_N), \tag{3.23}$$

where the rotation matrices $\mathbf{R}_x$, $\mathbf{R}_y$, $\mathbf{R}_z$ describe per definition rotations about space-fixed axes. The discussion of Eqs. (3.20) and (3.22) defines the sequence in which nested rotations have to be written down: the first rotation (i.e., the one on the right-hand side) has to be the rotation of the innermost axis, since this is the only rotation that does not affect the other ones. This can be formulated as

> **Rule 3**: For nested rotations, the sequence of rotations has to be written *from the inside out*, in order to ensure rotations about the correct axes.

Inserting Eqs. (3.14)–(3.16) into Eq. (3.23) leads to[4]

[4]The required work to find those matrices is much reduced using the *symbolic computation* packages offered by many scripting languages. For Python, the implementation is shown in Appendix C.4.2.

$$\mathbf{R}_{nautical} = \begin{bmatrix} \cos\theta_N \cos\phi_N, & \cos\theta_N \sin\phi_N \sin\psi_N - \sin\theta_N \cos\psi_N, & \cos\theta_N \sin\phi_N \cos\psi_N + \sin\theta_N \sin\psi_N \\ \sin\theta_N \cos\phi_N, & \sin\theta_N \sin\phi_N \sin\psi_N + \cos\theta_N \cos\psi_N, & \sin\theta_N \sin\phi_N \cos\psi_N - \cos\theta_N \sin\psi_N \\ -\sin\phi_N & \cos\phi_N \sin\psi_N & \cos\phi_N \cos\psi_N \end{bmatrix}. \tag{3.24}$$

This provides a convenient way to obtain the angles $(\theta_N, \phi_N, \psi_N)$ from the rotation matrix **R**

$$\begin{aligned} \phi_N &= -\arcsin(R_{zx}) \\ \theta_N &= \arcsin(\frac{R_{yx}}{\cos\phi_N}) \\ \psi_N &= \arcsin(\frac{R_{zy}}{\cos\phi_N}). \end{aligned} \tag{3.25}$$

Helmholtz Sequence

The nautical sequence is not the only sequence to describe the 3-D orientation of an object. Helmholtz (1867), another German physicist and physiologist from the nineteenth century, thought it would be better to start with a rotation about a horizontal axis. He characterized eye positions by a rotation about the horizontal interaural axis (i.e., the y-axis), followed by a rotation about the vertical axis, and then by a rotation about the line of sight, as shown in the right gimbal in Fig. 3.11:

$$\mathbf{R}_{Helm} = \mathbf{R}_y(\phi_H) \cdot \mathbf{R}_z(\theta_H) \cdot \mathbf{R}_x(\psi_H). \tag{3.26}$$

Fig. 3.11 In gimbal systems, the axes of rotation are determined by the geometry of system. Both gimbals in this figure are in the reference orientation. Let $\mathbf{b_x}$, $\mathbf{b_y}$, $\mathbf{b_z}$ describe a body-fixed coordinate system. **Left** In a *nautical (Fick) gimbal*, the orientation of the object (the turn on the inner dial) is completely characterized by a rotation about the vertical axis $\mathbf{b_z}$ by θ_N, followed by a rotation about the (rotated) horizontal axis $\mathbf{b_y}$ by ϕ_N, and a rotation about the (twice-rotated) dial-axis $\mathbf{b_x}$ by ψ_N. **Right** In a *Helmholtz gimbalh*, the orientation of the inner dial is characterized by a rotation first about the horizontal axis $\mathbf{b_y}$ by ϕ_H, followed by a rotation about the (rotated) $\mathbf{b_z}$ axis by θ_H, and then a rotation about the dial-axis $\mathbf{b_x}$ by ψ_H

The subscript "H" indicates that the angles refer to the *Helmholtz sequence* of rotations. One should keep in mind that the orientation of the object is characterized by the values of the rotation matrix $\mathbf{R}$, and $\mathbf{R}_{nautical}$ and $\mathbf{R}_{Helm}$ only give different sequence parameterizations for the rotation matrix. But once constructed, the matrix is used in the same manner. Using Eqs. (3.14)–(3.16) and matrix multiplication, we get

$$\mathbf{R}_{Helm} = \begin{bmatrix} \cos\theta_H \cos\phi_H & -\sin\theta_H \cos\phi_H \cos\psi_H + \sin\phi_H \sin\psi_H & \sin\theta_H \cos\phi_H \sin\psi_H + \sin\phi_H \cos\psi_H \\ \sin\theta_H & \cos\theta_H \cos\psi_H & -\cos\theta_H \sin\psi_H \\ -\cos\theta_H \sin\phi_H & \sin\theta_H \sin\phi_H \cos\psi_H + \cos\phi_H \sin\psi_H & -\sin\theta_H \sin\phi_H \sin\psi_H + \cos\phi_H \cos\psi_H \end{bmatrix} \tag{3.27}$$

When $\mathbf{R}$ is given, the Helmholtz angles $(\theta_H, \phi_H, \psi_H)$ can be using

$$\begin{aligned} \theta_H &= \arcsin(R_{yx}) \\ \phi_H &= -\arcsin\left(\frac{R_{zx}}{\cos\theta_H}\right) \\ \psi_H &= -\arcsin\left(\frac{R_{yz}}{\cos\theta_H}\right). \end{aligned} \tag{3.28}$$

Euler Sequence

Yet another sequence to describe 3-D orientation is common in theoretical physics and mechanics and in other technical literature, and often referred to as *Euler sequence*.[5]

In order to describe the movement of a spinning top rotating on a table, or of the earth during its rotation around the sun (see Fig. 3.12), three angles are needed: the intrinsic *rotation* (γ), *nutation* (β), and *precession* (α). Using these three angles, the orientation of the spinning object is described by (see Fig. 3.13)

- a rotation about the z-axis, by an angle α,
- followed by a rotation about the rotated x-axis, by an angle β, and
- followed by a rotation about the twice-rotated z-axis, by an angle γ.

$$\mathbf{R}_{Euler} = \mathbf{R}_z(\alpha) \cdot \mathbf{R}_x(\beta) \cdot \mathbf{R}_z(\gamma). \tag{3.29}$$

This leads to the parametrization

$$\mathbf{R}_{Euler} = \begin{bmatrix} -\sin\alpha_E \cos\beta_E \sin\gamma_E + \cos\alpha_E \cos\gamma_E & -\sin\alpha_E \cos\beta_E \cos\gamma_E - \cos\alpha_E \sin\gamma_E & \sin\alpha_E \sin\beta_E \\ \sin\alpha_E \cos\gamma_E + \cos\alpha_E \cos\beta_E \sin\gamma_E & -\sin\alpha_E \sin\gamma_E + \cos\alpha_E \cos\beta_E \cos\gamma_E & -\cos\alpha_E \sin\beta_E \\ \sin\beta_E \sin\gamma_E & \sin\beta_E \cos\gamma_E & \cos\beta_E \end{bmatrix} \tag{3.30}$$

[5]The expression *Euler angles* should be used very carefully: sometimes, these angles represent the *Euler sequence*, but often that expression is also applied when the nautical sequence is actually used!

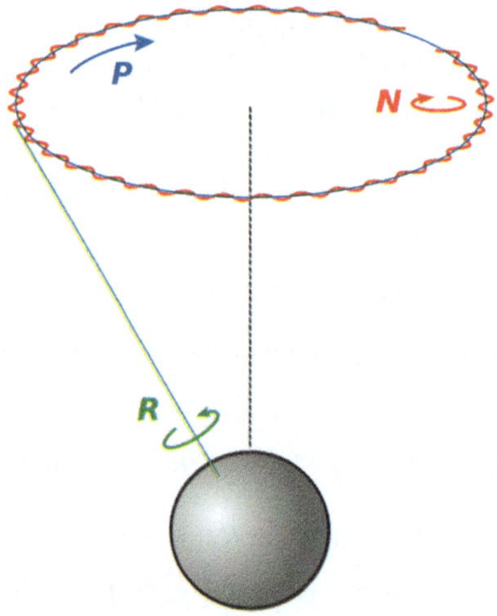

Fig. 3.12 Basic Euler motions of the earth. Intrinsic rotation (green "R"), precession (blue "P"), and nutation (red "N"). (From Wikipedia, original design by Dr. H. Sulzer)

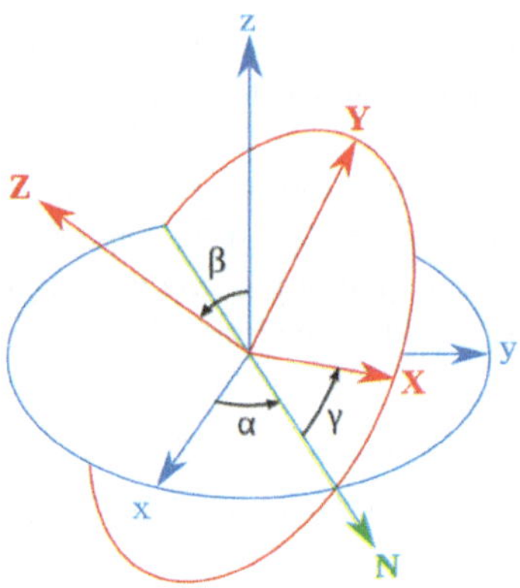

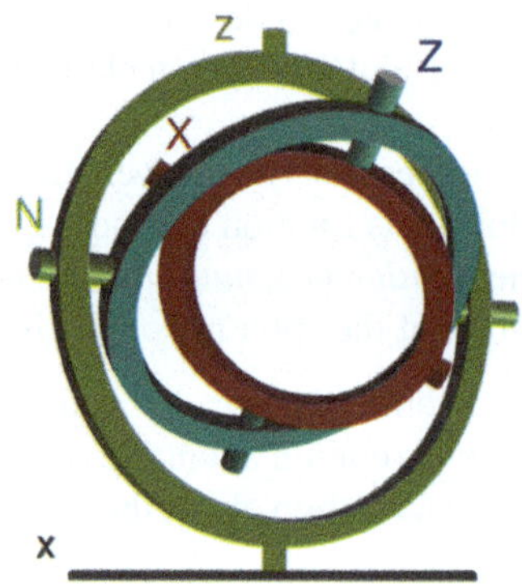

Fig. 3.13 Euler sequence: **Left**—The xyz (fixed) system is shown in blue, and the XYZ (rotated) system is shown in red. The line of nodes, labeled N, is shown in green (from Wikipedia). **Right** The corresponding gimbal

The corresponding angles can be calculated with

$$\begin{aligned} \gamma_E &= atan2(R_{zx}, R_{zy}) \\ \beta_E &= \arccos(R_{zz}) \\ \alpha_E &= -atan2(R_{xz}, R_{yz}). \end{aligned} \tag{3.31}$$

Notes:

- `atan2(a,b)` is equivalent to $\arctan(\frac{a}{b})$ where it also takes into account the quadrant that the point (a, b) is in.
- For β, there are generally two solutions in the interval $(\pi, \pi]$. The above formula works only when β is within the interval $[0, \pi)$.

Other Sequences

Depending on the application, sometimes yet other sequences can be used. For example, the most important angles in gait analysis are the knee and hip angles, which correspond approximately to rotations about the y-axis. Therefore, in gait analysis, the sequence $\mathbf{R}_z \cdot \mathbf{R}_x \cdot \mathbf{R}_y$ is common.

The names for different gimbal systems may differ significantly, depending on the area of application. In some contexts, angles of any type of gimbal system are referred to as "Euler angles". Angles from rotation sequences that involve all three axes *(x-y-z, y-z-x, z-x-y, x-z-y, z-y-x, y-x-z)* can be either called "Tait–Bryan angles", in honor of the Scottish mathematical physicist Peter Tait (1832–1901), who was—together with Hamilton—the leading exponent of quaternions, and the Welshman George Bryan (1864–1928), the originator of the equations of airplane motion or "Cardan angles", after the Renaissance mathematician, physician, astrologer, and gambler Jerome Cardan (1501–1576), who first described the cardan joint which can transmit rotary motion. And angles that have the same axis for the first and the last rotation (like the *Euler sequence* above) are called "proper Euler angles" *(z-x-z, x-y-x, y-z-y, z-y-z, x-z-x, y-x-y)*.

3.4.2 Gimbal Lock

Consider tracking a helicopter flying from the horizon toward an aerial gun, as indicated in Fig. 3.16. The helicopter flies toward the gun site and is tracked by the gun in elevation (Φ) and azimuth (Θ). When the helicopter is immediately above the gun site, the aerial gun is in the orientation indicated in Fig. 3.14. If the helicopter now changes direction and flies at 90° to its previous course, the gun cannot track this maneuver without a discontinuous jump in one or both of the gimbal orientations. There is no continuous motion that allows it to follow the target—it is in "gimbal lock". Note that even if the helicopter does not pass through the gimbal's zenith, but only near it, so that gimbal lock does not occur, the system must still move exceptionally rapidly to track the helicopter if it changed direction, as it rapidly passes from one bearing to the other. The closer to zenith the nearest point is, the faster this must be done, and if it actually goes through zenith, the limit of these "increasingly rapid" movements becomes infinitely fast, i.e., discontinuous. Another way to describe gimbal lock is to consider the inverse of matrices Eqs. (3.24, 3.27, and 3.30). The mathematical equivalent of a gimbal lock is if the calculation of the inverse of the matrices in Eqs. (3.24, 3.27, and 3.30) contains a divide by zero condition.

Fig. 3.14 Gimbal lock: the orientation of the innermost axis cannot be reoriented in the direction of the dotted arrows

3.5 Homogeneous Coordinates

3.5.1 Definition

An application where a tremendous amount of 3-D calculations have to be performed is the rendering of 3-D scenes in computer graphics. With every change in position and orientation of the observer the appearance of each 3-D surface element of the scene has to be re-calculated. The calculations require not only translations and rotations but also scaling and perspective distortions (McConnell 2005).

The mathematical discipline of *projective geometry* has found a way to perform all the required calculations efficiently in one step (see Fig. 3.15). For this approach to work, the number of coordinates for each point has to be increased from three to four, and the extended coordinates are called "homogeneous coordinates" or "projective coordinates":

$$\mathbf{p} \rightarrow \begin{pmatrix} \mathbf{p} \\ 1 \end{pmatrix}. \tag{3.32}$$

The additional, fourth element in the coordinates is essentially a scaling factor.

The matrix required to execute a generalized perspective transformation is a 4×4 matrix. For example, a rotation and translation can be "homogenized", i.e., executed in one step, as described in the following. A point $\mathbf{p}$ can be rotated and translated by

$$\mathbf{p}' = \mathbf{R} \cdot \mathbf{p} + \mathbf{t}, \tag{3.33}$$

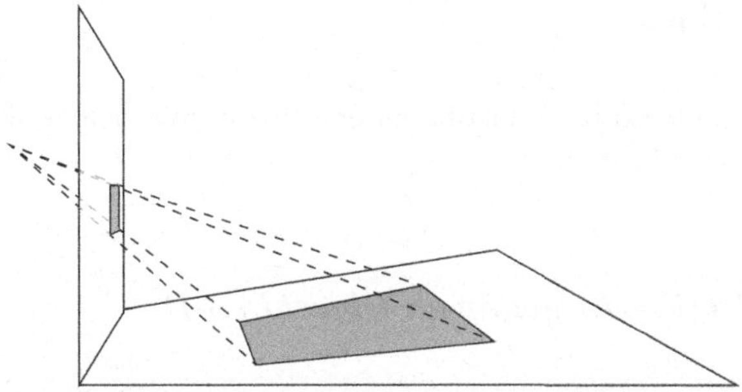

Fig. 3.15 Projection from one plane into another

where **p** indicates the starting point, **R** is the rotation matrix, **t** is the translation, and **p**′ is the new location. Using the "homogeneous" coordinates defined in Eq. (3.32), this can be written in one step as

$$\begin{pmatrix} \mathbf{p}' \\ 1 \end{pmatrix} = \begin{bmatrix} \mathbf{R} & \mathbf{t} \\ 0 & 1 \end{bmatrix} \cdot \begin{pmatrix} \mathbf{p} \\ 1 \end{pmatrix} = \begin{pmatrix} \mathbf{R} \cdot \mathbf{p} + \mathbf{t} \\ 1 \end{pmatrix}. \tag{3.34}$$

The resulting matrix $\begin{bmatrix} \mathbf{R} & \mathbf{t} \\ 0 & 1 \end{bmatrix}$ is a 4×4 matrix (since **R** is a 3×3 matrix, and one row and one column have been added) and is called "spatial transformation matrix". Similarly, geometric projections can be implemented by allowing general 4×4 matrices for the transformation.

By moving from 3-D Euclidian coordinates to 4-D homogeneous coordinates, all those transformations can be achieved with the same type of operation. The disadvantage: to represent a point, we now need four instead of the previous three numbers. The advantage: most image manipulations can now be performed in the same way. This homogeneous treatment of the different operations is reflected in the name, "homogeneous coordinates". This has become a huge advantage in computer vision: graphics cards can perform massive matrix multiplications. For comparison, modern multi-core CPUs (central processing units, i.e., the processor in a typical PC) have on the order of 16 pipelines, while GPUs (graphical processing units, the processors that are used for graphics cards) have on the order of 1024 pipelines. Thereby, instructions can be executed in parallel and can be dramatically optimized.

An important application of homogeneous coordinates are the "Denavit Hartenberg parameters". These are used in mechanical engineering to denote position and orientation of an end-link in robot manipulators (see Appendix B).

3.6 Applications

A few examples of the application of rotation matrices may help to show how to use them in practical applications.

3.6.1 *Two DOF—Targeting an Object in 3-D*

An aerial gun is mounted like a nautical gimbal: the outermost rotation is always about an earth-vertical axis (Fig. 3.16).

In the starting orientation ($\theta/\phi = 0/0$), the barrel of the aerial gun points straight ahead, i.e., along $\mathbf{b_x}$.

Task: When a target appears at $\mathbf{p} = (x, y, z)$, we want to reorient the gun such that the rotated gun barrel, which after the rotation points in the direction of $\mathbf{b}'_\mathbf{x}$, points at the target.

Solution: Taking the rotation matrix in the nautical sequence (Eq. 3.24), with $\psi = 0$ since a rotation about the line of the gun barrel is not relevant, we get

$$\mathbf{R}_{nautical}(\psi_N = 0) = \left[\mathbf{b}'_\mathbf{x}\ \mathbf{b}'_\mathbf{y}\ \mathbf{b}'_\mathbf{z}\right] = \begin{bmatrix} \cos\theta_N \cos\phi_N & -\sin\theta_N & \cos\theta_N \sin\phi_{N\,N} \\ \sin\theta_N \cos\phi_N & \cos\theta_N & \sin\theta_N \sin\phi_N \\ -\sin\phi_N & 0 & \cos\phi_N \end{bmatrix}.$$

With the first column $\mathbf{b}'_\mathbf{x} = \frac{\mathbf{p}}{|\mathbf{p}|}$, this leads to

$$\begin{aligned} \phi_N &= -\arcsin\left(\frac{p_z}{\sqrt{{p_x}^2+{p_y}^2+{P_z}^2}}\right) \\ \theta_N &= \arcsin\left(\frac{p_y}{\sqrt{{p_x}^2+{p_y}^2+{p_z}^2}} \cdot \frac{1}{\cos\phi_N}\right). \end{aligned} \tag{3.35}$$

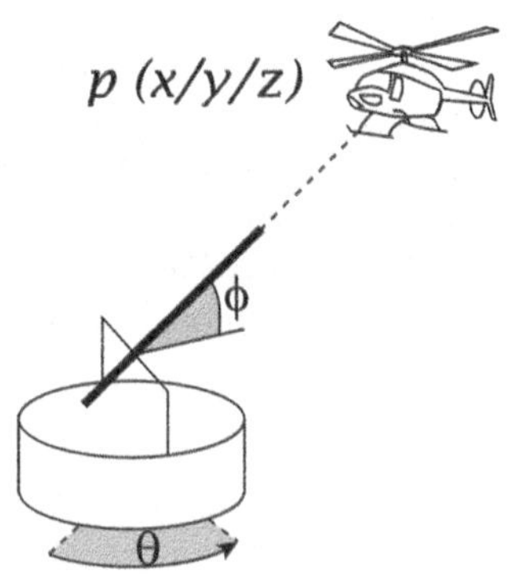

Fig. 3.16 Aerial Gun, tracking a target at $\mathbf{p} = (x/y/z)$

Note that a combination of a horizontal and a vertical rotation of the object in a well-defined sequence uniquely characterizes the direction of the forward direction. With eye movements, this is the line of sight, or gaze direction; with a gun turret on a ship this is the direction of the gun barrel (see Fig. 3.16). However, this does not completely determine the three-dimensional orientation of the object, since the rotation about the forward direction is still unspecified. A third rotation is needed to completely determine the orientation of the object. This third rotation ψ would not affect the direction in which the gun is pointing, it would only rotate around the pointing vector. For a quaternion solution to the targeting problem, see also Sect. 4.5.1.

3.6.2 *Two DOF—Projection onto a Flat Surface*

Another frequent paradigm is a projection onto a flat surface. Consider the following practical problem (Fig. 3.17).

Two projection systems are mounted at a distance d in front of a flat surface, and should both project a point at P, located on the screen at the location (hor/ver), where the positive horizontal direction on the screen is to the right, and the positive vertical direction up. The lower system is mounted like an aerial gun: it can rotate about a vertical axis (θ_N) and swivel about a (rotating) horizontal axis (ϕ_N). The system is mounted *below* cm lower than the center of the screen-based coordinate system, and when $\theta_N = \phi_N = 0$ the projector is pointing straight ahead toward the screen.

The upper projection system is mounted on a disk that hangs on the ceiling, *above* cm higher than the center of the screen-based coordinate system. It is mounted on a horizontal hinge which is parallel to the projection screen, so that it can rotate up and down (ϕ_H). The projector can swivel on the disk left/right (θ_H). And again, when $\theta_H = \phi_H = 0$ the projector is pointing straight ahead toward the screen.

Task: What are the projector angles for the lower projector (θ_N, ϕ_N), and for the upper projector (θ_H, ϕ_H), if both should point at the target $P = (hor/ver)$ on the screen?

Solution: The rotation sequence for the lower projector corresponds to the two outer rotations of a nautical gimbal, and the sequence of the upper projector to the two outer rotations of a Helmholtz gimbal, respectively (see also Fig. 3.11). In both cases, the rotation about the target direction is unimportant, and ψ in the equations for the nautical- and Helmholtz-rotation matrix can be set to zero. The direction to the target corresponds to the $\mathbf{b_x}$ axis after the rotation, and the target point is the intersection of this axis with a plane parallel to the $\mathbf{s_y}/\mathbf{s_z}$-plane at a distance d. So for the lower projector, the target is at

$$
\begin{aligned}
\mathbf{p} &= (d / -hor/ver + below) \\
\mathbf{b}'_{\mathbf{x}} &= \frac{\mathbf{p}}{|\mathbf{p}|}.
\end{aligned}
$$

(The sign before *hor* is negative, because the positive direction for "horizontal" on the screen is to the right, but the corresponding positive direction for "horizontal" for the projector is to the left.) From that the corresponding nautical angles can be determined with Eq. (3.35) from the previous example with the aerial gun. For the upper projector, the target is at

$$\mathbf{p} = (d/-hor/ver - above)$$
$$\mathbf{b}'_{\mathbf{x}} = \frac{\mathbf{p}}{|\mathbf{p}|}$$

and the projector angles can be found by applying the first two equations of Eq. (3.28):

$$\begin{aligned} \theta_H &= \arcsin\left(\frac{p_y}{\sqrt{{p_x}^2+{p_y}^2+{P_z}^2}}\right) \\ \phi_H &= -\arcsin\left(\frac{p_z}{\sqrt{{p_x}^2+{p_y}^2+{p_z}^2}} \cdot \frac{1}{\cos\theta_H}\right). \end{aligned} \quad (3.36)$$

A Python solution, with numbers approximating a setup such as Fig. 3.17, would be

```
from skinematics import vector
(d, hor, ver, above, below) = (1.5, 0.3, 0.2, 0.7, 1.4)
p_lower = [d, -hor, ver+below]
p_upper = [d, -hor, ver-above]
lower_projector = vector.target2orient(p_lower, orient_type='
    nautical')
upper_projector = vector.target2orient(p_upper, orient_type='
    Helmholtz')
```

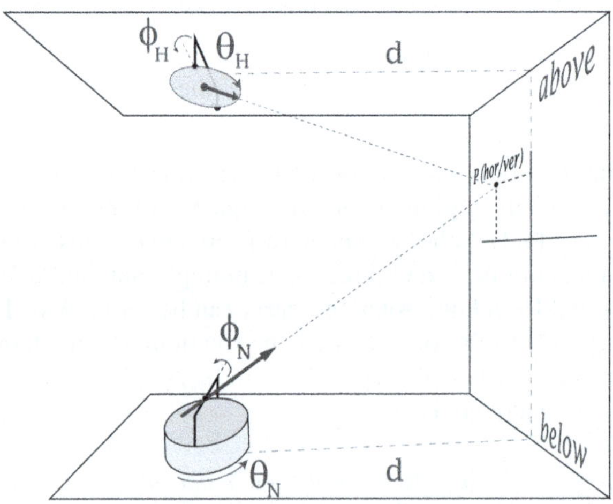

Fig. 3.17 Projection onto a flat surface, with different projection systems

Section 4.5.3 shows the solution of a somewhat more complex, but conceptually similar problem: orienting a camera in a missile such that it is directed on a selected target.

3.6.3 *Three DOF—3-D Orientation Measurements with Search Coils*

An interpretation of the values of the rotation matrix can be found by looking at Eq. (3.11): the columns of the rotation matrix **R** are equivalent to the vectors of the body-fixed coordinate system $\left[\mathbf{b_x}\ \mathbf{b_y}\ \mathbf{b_z}\right]$ expressed in the space-fixed coordinate system $\left[\mathbf{s_x}\ \mathbf{s_y}\ \mathbf{s_z}\right]$. Thus, for eye movement measurements with the search-coil method illustrated in Fig. 2.18, different values in the rotation matrix **R** indicate a different orientation of the eye-fixed coordinate system, i.e., a different orientation of the eye ball.

Task: What is the orientation of an eye on a gimbal that is rotated 15° to the left and 25° down, if it is (i) a nautical gimbal or (ii) a Helmholtz gimbal?

Solution: If an artificial eye ball on a nautical gimbal (Fig. 3.11a) is first turned 15° to the left and then (about the rotated axis $\mathbf{b_y}$) 25° down, i.e., $(\theta_N, \phi_N, \psi_N) = (15, 25, 0)$, its orientation will be given by the matrix

$$\mathbf{R}_{nautical} = \begin{bmatrix} 0.88 & -0.26 & 0.41 \\ 0.23 & 0.97 & 0.11 \\ -0.42 & 0 & 0.91 \end{bmatrix}.$$

Putting an eye on a Helmholtz gimbal (Fig. 3.11b), and turning it first 25° down and then 15° to the left (about the rotated axis $\mathbf{b_z}$), i.e., $(\phi_H, \theta_H, \psi_H) = (25, 15, 0)$, leads to a different orientation of the eye:

$$\mathbf{R}_{Helm} = \begin{bmatrix} 0.88 & -0.23 & 0.42 \\ 0.26 & 0.97 & 0 \\ -0.41 & 0.11 & 0.91 \end{bmatrix}.$$

The orientation of the eye in the two examples is clearly different: on the nautical gimbal $\mathbf{b_z}$ is given by (0.41, 0.11, 0.91), whereas on the Helmholtz gimbal it points in a different direction, (0.42, 0, 0.91).

Interpretation: Experimentally, the three-dimensional orientation of the eye in space can be measured, for example, with induction coils (see Figs. 2.17 and 2.18). When an induction coil is put into an oscillating magnetic field **m**, a voltage is induced in the coil (Robinson 1963). If the coil is characterized by a coil vector **c**, which is perpendicular to the coil and has a length proportional to the surface surrounded by the coil, the voltage induced is proportional to the cosine of the angle between **m** and **c** (Fig. 2.16). As pointed out by (Tweed et al. 1990), this leads to a simple correspondence between the values of the rotation matrix and the voltages induced in search coils. This connection can be demonstrated with the experimental setup shown in Fig. 2.18. Let

$$\mathbf{m_i} = \mathbf{s_i}\, m_i \, \sin(\omega_i \, t), \; i \in x, y, z$$

be three homogeneous orthogonal magnetic fields. They are parallel to the axes of the space-fixed coordinate system $\left[\mathbf{s_x}\, \mathbf{s_y}\, \mathbf{s_z}\right]$, have amplitudes m_i, and oscillate at frequencies ω_i. Further, let $\left[\mathbf{c_x}\, \mathbf{c_y}\, \mathbf{c_z}\right]$ denote three orthogonal coils parallel to the body-fixed coordinate system $\left[\mathbf{b_x}\, \mathbf{b_y}\, \mathbf{b_z}\right]$ and firmly attached to the object (here the eye). Then, the voltage induced by the magnetic field $\mathbf{m_i}$ in coil $\mathbf{c_j}$, V_{ij}, is given by

$$V_{ij} = R_{ij} * m_i * \omega_i * \cos(\omega_i * t) * c_j \text{ with } i, j \in x, y, z$$

where $c_j = \left|\mathbf{c_j}\right|$ indicates the length of the vector $\mathbf{c_j}$. This gives a direct interpretation of the elements of the rotation matrix $\mathbf{R}$: the voltage induced by the magnetic field $\mathbf{m_i}$ in the coil $\mathbf{c_j}$ is proportional to the element R_{ij} of the rotation matrix $\mathbf{R}$, which describes the rotation from the reference position, where the coils $\left[\mathbf{c_x}\, \mathbf{c_y}\, \mathbf{c_z}\right]$ line up with the magnetic fields $\left[\mathbf{m_x}\, \mathbf{m_y}\, \mathbf{m_z}\right]$, to the current position.

Note that for the determination of the 3-D orientation, three matrix elements suffice: H, V, and T indicate the signals that represent approximately the horizontal, vertical, and torsional orientation of the object:

$$\mathbf{R} = \begin{bmatrix} - & - & - \\ H & - & - \\ V & T & - \end{bmatrix}.$$

3.6.4 *Nested or Cascaded 3-D Rotation Sequences*

Nested or cascaded sequences are commonly found where one 3-D rotation follows another 3-D rotation. For example, the moving eye is placed inside a moving head, or a gimballed camera is mounted on a moving vehicle or missile. How can the formulas given above be used to derive the composite rotation of the two 3-D rotations combined? Furthermore, given the location of a point in space-fixed coordinates and eye coordinates, would it be possible to calculate the required rotation to ensure that the eye/camera looks at the point?

Camera on Moving Base

To describe the orientation of a camera-in-space (described by $\mathbf{R}_{camera}^{space}$), as shown in Fig. 3.18, one has to combine the orientation of the tilted base, e.g., a Google maps car (described by $\mathbf{R}_{base}^{space}$) and the orientation of the camera with respect to this base ($\mathbf{R}_{camera}^{base}$). To implement this mathematically, one has to use rotations of the coordinate system. This determines the sequence of the two rotations, and the rotation matrix describing the orientation of the camera-in-space is–according to the

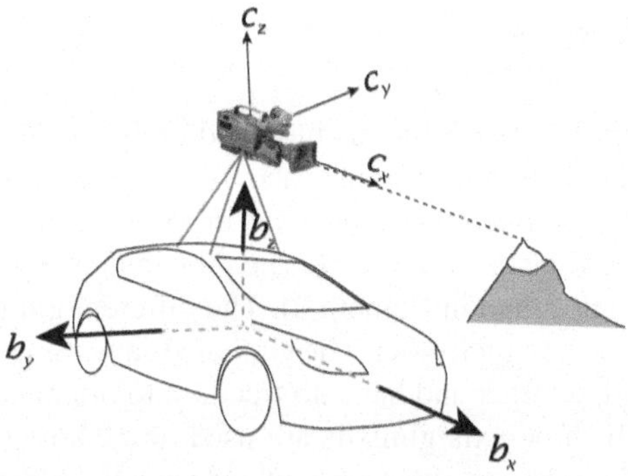

Fig. 3.18 A nested or cascaded rotation can, for example, be the orientation of a camera $[\mathbf{c}_x \mathbf{c}_y \mathbf{c}_z]$ on a moving base $[\mathbf{b}_x \mathbf{b}_y \mathbf{b}_z]$ fixating a target, here the top of a mountain

discussion following Eqs. (3.20) and (3.22)-given by

$$\mathbf{R}_{camera}^{space} = \mathbf{R}_{base}^{space} \cdot \mathbf{R}_{camera}^{base}. \tag{3.37}$$

From this, the orientation of the camera with respect to the base can be determined as

$$\mathbf{R}_{camera}^{base} = \left(\mathbf{R}_{base}^{space}\right)^{-1} \cdot \mathbf{R}_{camera}^{space}. \tag{3.38}$$

The way I personally remember these sequences: Take, for example, the following equation, which determines the orientation of the line of sight (LOS) of the camera

$$\mathbf{c}' = \mathbf{R}_{base}^{space} \cdot \left(\mathbf{R}_{camera}^{base} \cdot \mathbf{c}\right). \tag{3.39}$$

With base and camera in the reference orientation, $\mathbf{c}$ indicates the line of sight (LOS) of the camera. To find the current LOS, I first rotate the LOS of the camera on the base $\left(\mathbf{R}_{camera}^{base} \cdot \mathbf{c}\right)$. Then, in the second step I rotate the base, with the rotated camera already on it, to obtain the current orientation of the LOS: $\mathbf{R}_{base}^{space} \cdot \left(\mathbf{R}_{camera}^{base} \cdot \mathbf{c}\right)$.

Eye in Head

Similarly, let $\mathbf{R}_{head}^{space}$ be the rotation matrix describing the rotation of the reference frame with respect to a space-fixed coordinate system, and $\mathbf{R}_{eye}^{head}$ describe the rotation of the object in the reference frame (e.g., eye in head). Then

$$\mathbf{R}_{eye}^{space} = \mathbf{R}_{head}^{space} \cdot \mathbf{R}_{eye}^{head}, \tag{3.40}$$

and

$$\mathbf{R}_{eye}^{head} = \left(\mathbf{R}_{head}^{space}\right)^{-1} \cdot \mathbf{R}_{eye}^{space}. \tag{3.41}$$

3.6.5 *Camera Images*

Imagine the gaze vector of two cameras mounted on two different-sequenced two-axis gimbals following the same point in space. As the point moves around, the gaze vector of the two cameras traces the same locus in space. However, each camera will view differently rolled images of the world (i.e., the image "up" vector in Camera 1 will be at different angles than in Camera 2). The different gimbal's camera images will be rolled with respect to each other. The gimbaled cameras require all two degrees of freedom to follow the target and have no freedom to orientate their images to the upright orientation. If three-axis gimbals are used, there are no remaining degrees of freedom, and the images can always be rotated such that the required direction is pointing "up".

3.7 Exercises

Exercise 3.1: CT Scanner

A good example of a device that requires 3-D kinematics is a modern CT scanner.

Task: For the exercise, label the rotations about the three axes with α, β, and γ as shown in Fig. 3.19. With the CT scanner in the starting orientation, the outermost axis aligns with the z-axis, the middle axis with x, and the inner axis with y.

A patient has been attacked on the streets and has suffered from two gunshot wounds. When the patient is in a supine orientation (as indicated in Fig. 3.19), the first shot went through the left eye socket, into the direction $\overrightarrow{bullet_1} = [5\ 2\ 2]$; the second shot also went through the left eye socket, but in the direction $\overrightarrow{bullet_2} = [5\ -2\ -4]$.

Find out which settings for α, β, and γ bring the scanner into such an orientation that (i) $\overrightarrow{bullet_1}$ in the image is oriented along the y-axis of the scan and (ii) the trajectories of both bullets lie in the scanned plane (Fig. 3.20).

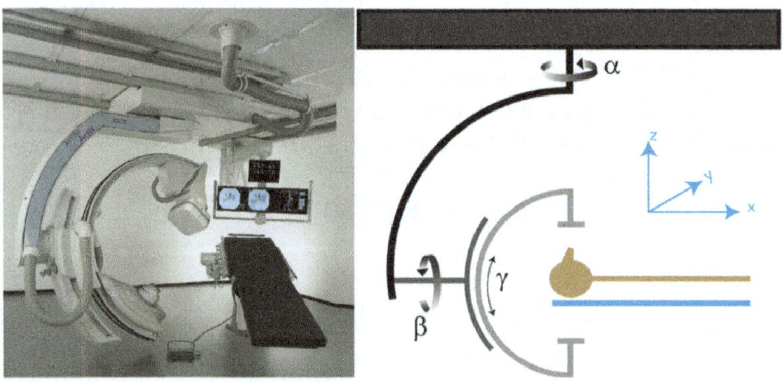

Fig. 3.19 (Left) An image of the Siemens Axiom Artis dTC scanner. (Right) Cartoon indicating the mathematical representation of this scanner

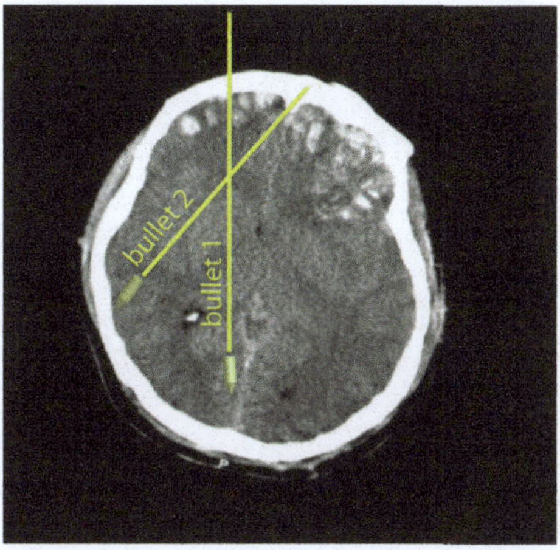

Fig. 3.20 CT scan of the head, with the bullet paths from two gunshot wounds schematically indicated. Note that the path of bullet 1 aligns with the y-axis

Exercise 3.2: Combining Rotation and Translation

This example will provide the first step in the measurement of movements with video-based systems. In the example, the movement of a comet that moves in a plane in space is observed with a camera, as shown in Fig. 3.22. The object is a comet that circles around a planet. The data units are 10^7 km. The tasks for this exercise are as follows:

- Read in the data from the file `planet_trajectory_2D.txt`, and write a program to calculate the planet velocity in the x- and y-directions. The data can be found in the *scikit-kinematics* package for Python users, and in the *Kinematics* toolbox for Matlab users, and are shown in Fig. 3.21.

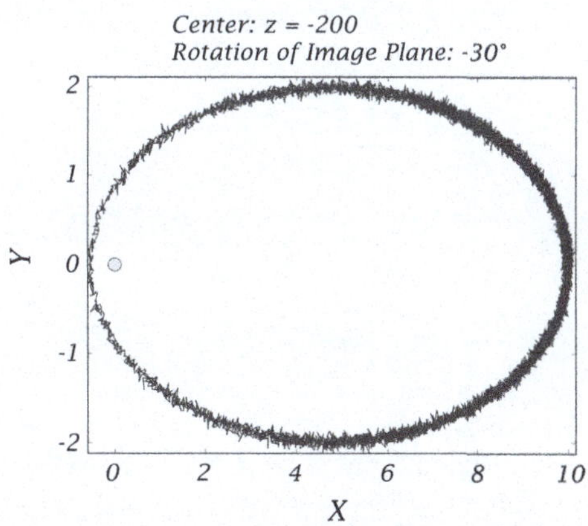

Fig. 3.21 Observed trajectory

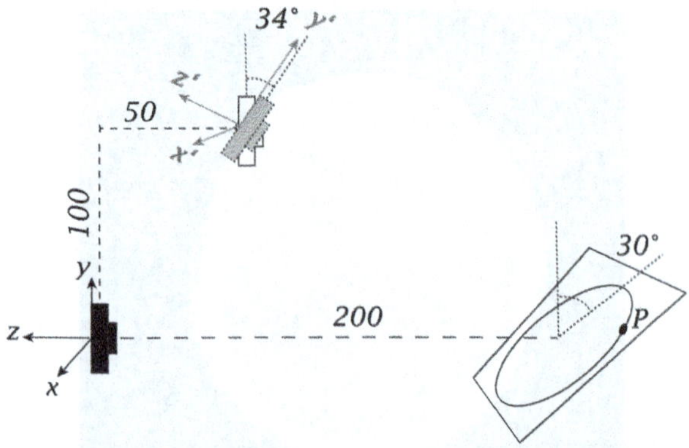

Fig. 3.22 Experimental setup: a particle moving in a tilted plane is observed from two different camera positions orientations

- External information is provided, which tells us that the data center is 200×10^7 km in front of the camera, and the trajectory lies in a plane that is tiled by 30° about the x-axis. Calculate the 3-D position of the trajectory of the particle (see Fig. 3.22).
- Calculate the 3-D position $p_{shifted}$ of the comet in camera coordinates, observed from a satellite which has traveled 50×10^7 km toward the planet, and 100×10^7 km orthogonally to it.
- Calculate the position $p_{shiftRot}$ (with respect to the camera), if the satellite is rotated 34° downward.

python **Code:** `C3_examples_rotmat.py`: Python examples of different operations with rotation matrices, such as generating symbolic and numeric rotation matrices and calculating the corresponding rotation sequences. (p. 141)

Chapter 4
Quaternions and Gibbs Vectors

While most readers are familiar with the rotation matrices presented in the previous chapter, rotation matrices are not the most convenient or efficient way to represent rotations. Euler had already realized that expressing a rotation with a vector parallel to the axis of rotation was more elegant than using a rotation matrix. And the mathematical work by Hamilton and Gibbs on alternative representations of rotations, which is presented in this chapter, prepared the way not only for an efficient representation of rotations, but for the whole modern vector calculus.

4.1 Representing Rotations by Vectors

Rotation matrices are not the most efficient way to describe a rotation: they have nine elements, yet only three are actually needed to uniquely characterize a rotation. Another disadvantage of describing 3-D rotations with rotation matrices is that the three axes of rotation, as well as the sequence of the rotations about these axes, have to be defined arbitrarily, with different sequences leading to different rotation angles. Euler's rotational theorem (Euler 1775) states that a more efficient way to characterize a rotation is to use a vector: the axis of rotation is defined by the direction of the vector $\mathbf{q}$, and the rotation magnitude θ is defined by the vector length (Fig. 4.1). The orientation is defined by the right-hand-rule (Fig. 4.2). Such a vector has only three parameters, and no sequence of multiple rotations is involved.

Different conventions can be used to define the vector:

"Euler vectors"	$\|\mathbf{q}\| = \theta$	Sect. 4.2
"Quaternions"	$\|\mathbf{q}\| = \sin(\theta/2)$	Sect. 4.3
"Gibbs vectors"	$\|\mathbf{q}\| = \tan(\theta/2)$	Sect. 4.4

Rotation matrices are often an easy way to establish a correspondence between measured values (e.g., induction coil voltages, or images) and the orientation of an object relative to a given reference orientation. But for working with 3-D orientations and for calculations, quaternions or Gibbs vectors have proven to be more

T. Haslwanter, *3D Kinematics*, https://doi.org/10.1007/978-3-319-75277-8_4

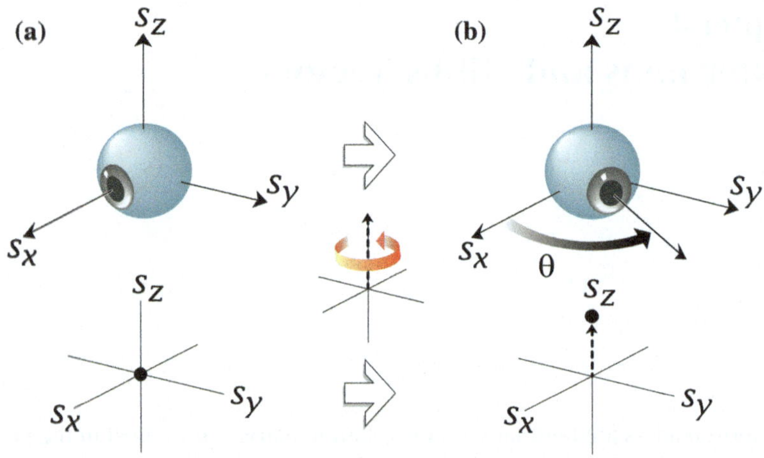

Fig. 4.1 Description of 3-D eye orientation by a vector: **a** The eye in the reference orientation (top) corresponds to the zero vector (bottom). **b** A different horizontal eye orientation (top) can be reached by rotating the eye from the reference orientation about the $\mathbf{s_z}$ -axis. This eye orientation is, thus, represented by a vector along the $\mathbf{s_z}$-axis, with a length proportional to the angle of the rotation (bottom). Note that usually only the end-point of the vector describing the eye orientation is shown, not the whole vector

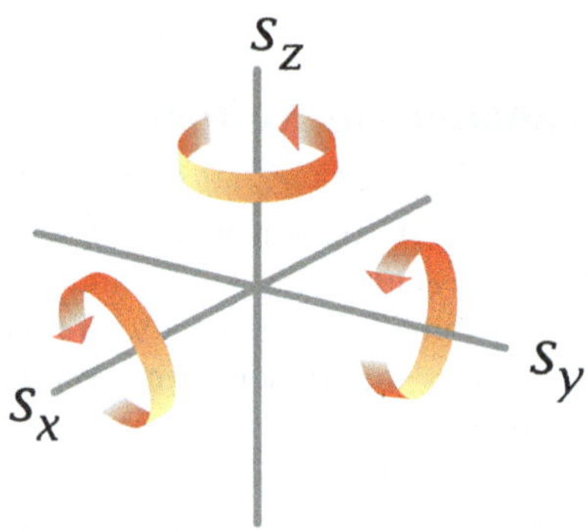

Fig. 4.2 According to the right-hand-rule *positive* rotations are yaw-rotations to the left (about $\mathbf{s_z}$), pitch-rotations downward (about $\mathbf{s_y}$), and roll-rotations clockwise as seen from the object (about $\mathbf{s_x}$)

intuitive and efficient. They are nonredundant, using three parameters to describe the three degrees of freedom of rotations. And they do not require an arbitrarily chosen sequence of rotations, but describe orientation by a single rotation from the reference orientation to the current orientation. In addition, they form an intuitive way of parameterizing rotations by expressing them by their axis and size, allow for an easy combination of rotations, and are more accurate when used to integrate incremental changes in orientation over time.

4.2 Axis-Angle Euler Vectors

A vector $\mathbf{x}$ can be rotated by an angle ρ about a vector $\mathbf{n}$ through

$$\mathbf{R}(\mathbf{n}, \rho) \cdot \mathbf{x} = (\mathbf{n} \cdot \mathbf{x}) * \mathbf{n} + \mathbf{n} \times \mathbf{x} * \sin(\rho) - \mathbf{n} \times (\mathbf{n} \times \mathbf{x}) \cos(\rho) \tag{4.1}$$

or equivalently

$$\mathbf{R}(\mathbf{n}, \rho) \cdot \mathbf{x} = \mathbf{x} * \cos(\rho) + (1 - \cos(\rho)) * (\mathbf{n} \cdot \mathbf{x}) * \mathbf{n} + \sin(\rho) * \mathbf{n} \times \mathbf{x}\,. \tag{4.2}$$

The development of this parametrization of rotations can probably be attributed to (Rodrigues 1840), and Eq. (4.2) is, therefore, also called "Rodrigues' rotation formula". The representation of a rotation with an axis $\mathbf{n}$ and an angle ρ is sometimes referred to as "axis-angle representation" of a rotation. And in honor of Euler's rotation theorem (see p. 179), a vector with a direction $\mathbf{n}$ and a length ρ is called "Euler vector".

Note: While Euler vectors give a convenient representation of a rotation, no equation exists that allows to combine two Euler vectors. Therefore, practical implementations of rotations have to be based on rotation matrices, quaternions or Gibbs vectors (see below).

4.3 Quaternions

4.3.1 Background

The theory of quaternions was invented and developed by Hamilton in the mid-nineteenth century (Hamilton 1844). Hamilton realized that the complex numbers could be interpreted as points in a plane (see Fig. 3.3), and he was looking for a way to do the same for points in three-dimensional space. Points in space can be represented by their coordinates, which are triples of numbers. For many years he had known how to add and subtract triples of numbers. However, Hamilton had been stuck on the problem of multiplication and division for a long time. He could not figure out how to calculate the quotient of the coordinates of two points in space. Hamilton found that he could not accomplish this by using 3-component vectors, but had to use 4 components. He called these quadruples "quaternions".

A detailed treatment of quaternions and their elegant mathematical properties can be found in mathematical texts (Brand 1948; Altmann 1986; Kuipers 1999), many papers on eye movements (Westheimer 1957; Tweed and Vilis 1987; Hepp et al. 1989; Tweed et al. 1990), and papers in more technical journals (Rooney 1977; Funda and Paul 1988). Recommendable introductions are also available on the Internet (see Appendix G).

A note for physicists, or for the mathematically more curious reader: quaternions are four-dimensional representations of *Clifford algebras* (see also Appendix A.4). The two-dimensional representations are the complex 2 × 2-matrices, or "Pauli spin matrices" (SU2, or two-dimensional special unitary group).And the three-dimensional representations are the rotation matrices (SO3, or special three-dimensional orthogonal group). Especially in theoretical physics, the advantages of switching from 3-D representations over to 4-D quaternions can be massive (Girard 1984). For example, using Clifford algebra, the four Maxwell equations can be written in just one very compact, elegant equation (see also Appendix A.4):

$$\nabla F = \mu_0 J\,. \tag{4.3}$$

4.3.2 Quaternion Properties

The following description of quaternions will cover only their essential properties, and no mathematical proofs will be given.

A full quaternion $\tilde{\boldsymbol{q}}$ has four components, and is given by

$$\tilde{\boldsymbol{q}} = q_0 + (q_1 * \tilde{\boldsymbol{i}} + q_2 * \tilde{\boldsymbol{j}} + q_3 * \tilde{\boldsymbol{k}}) = q_0 + \mathbf{q} \cdot \mathbf{I}, \tag{4.4}$$

where $\mathbf{q} = \begin{pmatrix} q_1 \\ q_2 \\ q_3 \end{pmatrix}$, $\mathbf{I} = \begin{pmatrix} \tilde{\boldsymbol{i}} \\ \tilde{\boldsymbol{j}} \\ \tilde{\boldsymbol{k}} \end{pmatrix}$, and $(\tilde{\boldsymbol{i}}, \tilde{\boldsymbol{j}}, \tilde{\boldsymbol{k}})$ are defined by

$$\begin{array}{lll} \tilde{\boldsymbol{i}} \cdot \tilde{\boldsymbol{i}} = -1 & \tilde{\boldsymbol{j}} \cdot \tilde{\boldsymbol{j}} = -1 & \tilde{\boldsymbol{k}} \cdot \tilde{\boldsymbol{k}} = -1 \\ \tilde{\boldsymbol{i}} \cdot \tilde{\boldsymbol{j}} = \tilde{\boldsymbol{k}} & \tilde{\boldsymbol{j}} \cdot \tilde{\boldsymbol{k}} = \tilde{\boldsymbol{i}} & \tilde{\boldsymbol{k}} \cdot \tilde{\boldsymbol{i}} = \tilde{\boldsymbol{j}} \\ \tilde{\boldsymbol{j}} \cdot \tilde{\boldsymbol{i}} = -\tilde{\boldsymbol{k}} & \tilde{\boldsymbol{k}} \cdot \tilde{\boldsymbol{j}} = -\tilde{\boldsymbol{i}} & \tilde{\boldsymbol{i}} \cdot \tilde{\boldsymbol{k}} = -\tilde{\boldsymbol{j}} \end{array} \tag{4.5}$$

q_0 is often called the "scalar component" of the quaternion $\tilde{\boldsymbol{q}}$, and $\mathbf{q}$ the "vector component" of $\tilde{\boldsymbol{q}}$. (Note that the quaternion is written as $\tilde{\boldsymbol{q}}$, and the quaternion *vector* as $\mathbf{q}$.)

With Eq. (4.5), one can show (see Appendix A.3) that the multiplication of two quaternions $\tilde{\boldsymbol{p}}$ and $\tilde{\boldsymbol{q}}$, here denoted "∘", is given by

$$\tilde{\boldsymbol{q}} \circ \tilde{\boldsymbol{p}} = \sum_{i=0}^{3} q_i\, I_i * \sum_{j=0}^{3} p_j\, I_j \;=\; (q_0 p_0 - \mathbf{q} \cdot \mathbf{p}) + (q_0 \mathbf{p} + p_0 \mathbf{q} + \mathbf{q} \times \mathbf{p}) \cdot \mathbf{I}\,. \tag{4.6}$$

The right side of Eq. (4.6) is obtained by using the definitions in Eqs. (4.4) and (4.5). Similar to rotation matrices, the sequence of the quaternions in Eq. (4.6) is important, and the opposite sequence, $\tilde{\boldsymbol{p}} \circ \tilde{\boldsymbol{q}}$, would lead to a different quaternion.

The *inverse quaternion* is given by

$$\tilde{q}^{-1} = \frac{q_0 - \mathbf{q} \cdot \mathbf{I}}{|q|^2} . \tag{4.7}$$

The norm of a quaternion is given by the quadrature sum of all four components

$$|\tilde{q}| = \sqrt{\sum_{i=0}^{3} q_i^2} . \tag{4.8}$$

4.3.3 Interpretation of Quaternions

To interpret quaternions, it is helpful to group them into four classes:

(1) Quaternions with the scalar component equal to 0 correspond to $\mathbb{R}^3$, the space of three-dimensional vectors. (This group is sometimes also called "pure quaternions".)
(2) Quaternions with a zero vector component $\mathbf{0}$ correspond to the space of scalars, $\mathbb{R}$.
(3) Unit quaternions, i.e., quaternions with $|\tilde{q}| = 1$, correspond to $SO3$, the group of orthogonal matrices with a determinant of 1. Unit quaternions, sometimes also called "rotation quaternions", can be used to describe rotations in space.
(4) General quaternions with scalar and vector components unequal zero, with a norm unequal to 1. These quaternions describe a combination of a rotation and scaling of vectors (Rooney 1977). If the norm of the quaternion is > 1, the objects are stretched; and if the norm is < 1, objects are compressed.

4.3.4 Unit Quaternions

A quaternion describing a pure rotation in 3-D space is a "unit quaternion" and has a norm of $|\tilde{q}| = 1$.

From Eq. (4.7), the inverse quaternion $\tilde{q}^{-1}$ for a unit quaternion is given by

$$\tilde{q}^{-1} = q_0 - \mathbf{q} \cdot \mathbf{I} , \tag{4.9}$$

Comparing Eqs. (4.5) and (3.9) to Fig. 3.3, which describes rotations in the complex plane, one can find a physical interpretation for $\tilde{i}$, $\tilde{j}$, and $\tilde{k}$. A rotation of a complex number c by an angle ϕ is given by $c' = e^{j\phi} \cdot c$, where $j * j = -1$, and j can be interpreted as a vector pointing perpendicular out of the 2-D-plane. To describe rotations in 3-D, we need three axes to rotate about: $\tilde{\boldsymbol{i}}$, $\tilde{\boldsymbol{j}}$, and $\tilde{\boldsymbol{k}}$. It can be shown that for a quaternion of the form

$$\tilde{\boldsymbol{q}} = \begin{pmatrix} 0 \\ \theta/2 * \mathbf{v} \end{pmatrix} \tag{4.10}$$

where $|\mathbf{v}| = 1$, the exponential of the quaternion is given by the unit quaternion

$$\exp(\tilde{\boldsymbol{q}}) = \begin{pmatrix} \cos(\theta/2) \\ \sin(\theta/2) * \mathbf{v} \end{pmatrix} \tag{4.11}$$

generalizing Eq. (3.5).

A unit quaternion describes a rotation by an angle θ around an axis described by the unit vector $\mathbf{n} = (n_i \tilde{\boldsymbol{i}}, n_j \tilde{\boldsymbol{j}}, n_k \tilde{\boldsymbol{k}})$

$$\tilde{\boldsymbol{q}} = \cos(\theta/2) + \sin(\theta/2)[n_i \tilde{\boldsymbol{i}} + n_j \tilde{\boldsymbol{j}} + n_k \tilde{\boldsymbol{k}}] = q_0 + \mathbf{q} \cdot \mathbf{I}, \tag{4.12}$$

where the orientation of $\mathbf{n}$ describes the axis of rotation, as shown in Fig. 4.1b. The length of the vector component equals $\sin(\theta/2)$. The unit quaternion has the following properties:

$$|\tilde{\boldsymbol{q}}| = \sqrt{\cos^2(\theta/2) + \sin^2(\theta/2)} = 1 \tag{4.13a}$$

$$|\mathbf{q}| = \sqrt{q_1^2 + q_2^2 + q_3^2} = \sin(\theta/2) \tag{4.13b}$$

$$\mathbf{q} \parallel \mathbf{n} \tag{4.13c}$$

The $\theta/2$ property of rotation quaternions, i.e. the fact that the lengths of a unit quaternion vector is determined by *half* the rotation angle, $\theta/2$, can be explained by interpreting a rotation as a sum of two reflections, see Appendix Fig. A.4. Another way to explain it is by considering Eq. (4.14) discussed below. The rotation quaternion appears twice. This "double application" of $\theta/2$ leads to a final rotation by an angle θ.

Examples

40° yaw movement to the left: A yaw movement is a rotation about a vertical axis, so the quaternion vector has to be along the axis $\mathbf{n} = \begin{pmatrix} 0 \\ 0 \\ 1 \end{pmatrix}$. The yaw rotation to the left is positive (see Fig. 4.2). And since the magnitude of the rotation is 40°, the full quaternion is

$$\tilde{\boldsymbol{q}} = \begin{pmatrix} q_0 \\ \mathbf{q} \end{pmatrix} = \begin{pmatrix} \cos(40°/2) \\ 0 \\ 0 \\ \sin(40°/2) \end{pmatrix} = \begin{pmatrix} \cos(20°) \\ 0 \\ 0 \\ \sin(20°) \end{pmatrix}.$$

90° pitch rotation nose-up: A pitch movement is a rotation about the *y-axis*, so the quaternion vector has to be along the axis $\mathbf{n} = \begin{pmatrix} 0 \\ 1 \\ 0 \end{pmatrix}$. A pitch rotation nose-up is negative (Fig. 4.2). And since the magnitude of the rotation is 90°, the full quaternion is $\tilde{\boldsymbol{q}} = \begin{pmatrix} \cos(-45°) \\ 0 \\ \sin(-45°) \\ 0 \end{pmatrix} = \begin{pmatrix} \cos(45°) \\ 0 \\ -\sin(45°) \\ 0 \end{pmatrix}$.

Relation to Rotation Matrix

The connection between a rotation quaternion $\tilde{\boldsymbol{q}}$ and a rotation matrix $\mathbf{R}$, both describing the rotation of a vector $\mathbf{x}$ about an axis $\mathbf{n}$ by an angle θ, can be derived from the definition of quaternions in Eqs. (4.4)–(4.13):

$$\begin{aligned} \tilde{\boldsymbol{x}}' &= \tilde{\boldsymbol{q}} \circ \tilde{\boldsymbol{x}} \circ \tilde{\boldsymbol{q}}^{-1} = \begin{pmatrix} 0 \\ \mathbf{x}' \end{pmatrix} \\ \mathbf{x}' &= \mathbf{R} \cdot \mathbf{x}\,. \end{aligned} \tag{4.14}$$

The proof of Eq. (4.14) is given in Appendix A.3.

$\tilde{\boldsymbol{x}}'$ in Eq. (4.14) is a full quaternion, but the scalar component evaluates to zero $q_0 = 0$. Hence, the rotation matrix $\mathbf{R}$ corresponding to the quaternion $\tilde{\boldsymbol{q}}$ can be determined as

$$\mathbf{R} = \begin{bmatrix} q_0^2 + q_1^2 - q_2^2 - q_3^2 & 2(q_1q_2 - q_0q_3) & 2(q_1q_3 + q_0q_2) \\ 2(q_1q_2 + q_0q_3) & q_0^2 - q_1^2 + q_2^2 - q_3^2 & 2(q_2q_3 - q_0q_1) \\ 2(q_1q_3 - q_0q_2) & 2(q_2q_3 + q_0q_1) & q_0^2 - q_1^2 - q_2^2 + q_3^2 \end{bmatrix}. \tag{4.15}$$

The inverse computation is given by

$$\mathbf{q} = 0.5 * copysign \begin{pmatrix} \sqrt{1 + R_{xx} - R_{yy} - R_{zz}}, & R_{zy} - R_{yz} \\ \sqrt{1 - R_{xx} + R_{yy} - R_{zz}}, & R_{xz} - R_{zx} \\ \sqrt{1 - R_{xx} - R_{yy} + R_{zz}}, & R_{yx} - R_{xy} \end{pmatrix}, \tag{4.16}$$

where $copysign(x, y) = sign(y) * |x|$.

Sequential Rotations with Quaternions

Equation (4.14) is the quaternion equivalent of a matrix multiplication for rotation matrices $\mathbf{R}$. Therefore, a sequence of quaternion rotations is the same as the sequence of the corresponding rotation matrices. For combined rotations, care has to be taken with the sequence of quaternions: the same rules apply as for rotation matrices, which means that the first rotation acting on a vector is on the right-hand side of the quaternion multiplication in Eq. (4.6).

Relation to Sequential Rotations

The examples above show how quaternions are related to rotations about coordinate axes. Using the rules for quaternion multiplication Eq. (4.6), one can calculate the relationship between rotation angles for the nautical sequence and quaternions:

$$\tilde{q_z}(\theta_N) \circ \tilde{q_y}(\phi_N) \circ \tilde{q_x}(\psi_N) = \begin{pmatrix} \cos\frac{\theta_N}{2} * \cos\frac{\phi_N}{2} * \cos\frac{\psi_N}{2} + \sin\frac{\theta_N}{2} * \sin\frac{\phi_N}{2} * \sin\frac{\psi_N}{2} \\ \cos\frac{\theta_N}{2} * \cos\frac{\phi_N}{2} * \sin\frac{\psi_N}{2} - \sin\frac{\theta_N}{2} * \sin\frac{\phi_N}{2} * \cos\frac{\psi_N}{2} \\ \cos\frac{\theta_N}{2} * \sin\frac{\phi_N}{2} * \cos\frac{\psi_N}{2} + \sin\frac{\theta_N}{2} * \cos\frac{\phi_N}{2} * \sin\frac{\psi_N}{2} \\ \sin\frac{\theta_N}{2} * \cos\frac{\phi_N}{2} * \cos\frac{\psi_N}{2} - \cos\frac{\theta_N}{2} * \sin\frac{\phi_N}{2} * \sin\frac{\psi_N}{2} \end{pmatrix} . \tag{4.17}$$

For the Helmholtz sequence, this leads to

$$\tilde{q_y}(\phi_H) \circ \tilde{q_z}(\theta_H) \circ \tilde{q_x}(\psi_H) = \begin{pmatrix} \cos\frac{\theta_H}{2} * \cos\frac{\phi_H}{2} * \cos\frac{\psi_H}{2} - \sin\frac{\theta_H}{2} * \sin\frac{\phi_H}{2} * \sin\frac{\psi_H}{2} \\ \cos\frac{\theta_H}{2} * \cos\frac{\phi_H}{2} * \sin\frac{\psi_H}{2} + \sin\frac{\theta_H}{2} * \sin\frac{\phi_H}{2} * \cos\frac{\psi_H}{2} \\ \cos\frac{\theta_H}{2} * \sin\frac{\phi_H}{2} * \cos\frac{\psi_H}{2} + \sin\frac{\theta_H}{2} * \cos\frac{\phi_H}{2} * \sin\frac{\psi_H}{2} \\ \sin\frac{\theta_H}{2} * \cos\frac{\phi_H}{2} * \cos\frac{\psi_H}{2} - \cos\frac{\theta_H}{2} * \sin\frac{\phi_H}{2} * \sin\frac{\psi_H}{2} \end{pmatrix} . \tag{4.18}$$

And for the Euler sequence we get

$$\tilde{q_z}(\alpha_E) \circ \tilde{q_x}(\beta_E) \circ \tilde{q_z}(\gamma_E) = \begin{pmatrix} \cos\frac{\alpha_E}{2} * \cos\frac{\beta_E}{2} * \cos\frac{\gamma_E}{2} - \sin\frac{\alpha_E}{2} * \cos\frac{\beta_E}{2} * \sin\frac{\gamma_E}{2} \\ \cos\frac{\alpha_E}{2} * \sin\frac{\beta_E}{2} * \cos\frac{\gamma_E}{2} + \sin\frac{\alpha_E}{2} * \sin\frac{\beta_E}{2} * \sin\frac{\gamma_E}{2} \\ \sin\frac{\alpha_E}{2} * \sin\frac{\beta_E}{2} * \cos\frac{\gamma_E}{2} - \cos\frac{\alpha_E}{2} * \sin\frac{\beta_E}{2} * \sin\frac{\gamma_E}{2} \\ \cos\frac{\alpha_E}{2} * \cos\frac{\beta_E}{2} * \sin\frac{\gamma_E}{2} + \sin\frac{\alpha_E}{2} * \cos\frac{\beta_E}{2} * \cos\frac{\gamma_E}{2} \end{pmatrix} . \tag{4.19}$$

The inverse relationships, i.e., calculating the angles for the different rotation sequences, can be obtained by inserting the corresponding matrix elements from Eq. (4.15) into Eq. (3.24) for the nautical sequence, Eq. (3.27) for the Helmholtz sequence, and Eq. (3.30) for the Euler sequence.

4.4 Gibbs Vectors

4.4.1 Properties of Gibbs Vectors

Gibbs vectors are named after Josiah Willard Gibbs (1839–1903), the inventor of—among many other things—modern vector calculus. A *Gibbs vector*[1] **r** corresponding to the rotation matrix **R** is given by

$$\mathbf{r} = \frac{1}{1 + (R_{11} + R_{22} + R_{33})} * \begin{pmatrix} R_{32} - R_{23} \\ R_{13} - R_{31} \\ R_{21} - R_{12} \end{pmatrix} . \tag{4.20}$$

From this one can show that

$$|\mathbf{r}| = \tan(\rho/2) . \tag{4.21}$$

[1]Some authors call Gibbs vectors "rotation vectors".

The coefficients of the Gibbs vectors are sometimes referred to as "Rodrigues parameters" (Altmann 1986; Dai 2015).

Finding the relationship between Gibbs vectors and other descriptions of rotations, such as nautical angles, requires an equation for combined rotations with Gibbs vectors. Using Eqs. (4.6) and (4.21) gives

$$\mathbf{r_q} \circ \mathbf{r_p} = \frac{\mathbf{r_q} + \mathbf{r_p} + \mathbf{r_q} \times \mathbf{r_p}}{1 - \mathbf{r_q} \cdot \mathbf{r_p}}, \tag{4.22}$$

where $\mathbf{r_p}$ is the first rotation (about a space-fixed axis parallel to $\mathbf{r_p}$), and $\mathbf{r_q}$ the second rotation (about a space-fixed axis parallel to $\mathbf{r_q}$).

The Gibbs vector corresponding to the nautical angles in Eq. (3.23) can be obtained by combining three Gibbs vectors with Eq. (4.22). Denoting a Gibbs vector which describes a rotation about an axis $\mathbf{n}$ by an angle θ with $\mathbf{r}\,(\mathbf{n}, \theta)$, this leads to

$$\begin{aligned} \mathbf{r} &= \mathbf{r}(\mathbf{e_3}, \theta_N) \circ \mathbf{r}(\mathbf{e_y}, \phi_N) \circ \mathbf{r}(\mathbf{e_1}, \psi_N) = \\ &= \frac{1}{1 - \tan(\frac{\theta_N}{2}) * \tan(\frac{\phi_N}{2}) * \tan(\frac{\psi_N}{2})} \begin{pmatrix} \tan\frac{\psi_N}{2} & - & \tan\frac{\theta_N}{2} * \tan\frac{\phi_N}{2} \\ \tan\frac{\phi_N}{2} & + & \tan\frac{\theta_N}{2} * \tan\frac{\psi_N}{2} \\ \tan\frac{\theta_N}{2} & - & \tan\frac{\phi_N}{2} * \tan\frac{\psi_N}{2} \end{pmatrix}, \end{aligned} \tag{4.23}$$

where $(\theta_A, \phi_A, \psi_A)$ are the nautical angles. For Helmholtz angles, the corresponding equation reads

$$\begin{aligned} \mathbf{r} &= \mathbf{r}(\mathbf{e_y}, \phi_H) \circ \mathbf{r}(\mathbf{e_3}, \theta_H) \circ \mathbf{r}(\mathbf{e_1}, \psi_H) = \\ &= \frac{1}{1 - \tan(\frac{\theta_H}{2}) * \tan(\frac{\phi_H}{2}) * \tan(\frac{\psi_H}{2})} \begin{pmatrix} \tan\frac{\psi_H}{2} & + & \tan\frac{\theta_H}{2} * \tan\frac{\phi_H}{2} \\ \tan\frac{\phi_H}{2} & + & \tan\frac{\theta_H}{2} * \tan\frac{\psi_H}{2} \\ \tan\frac{\theta_H}{2} & - & \tan\frac{\phi_H}{2} * \tan\frac{\psi_H}{2} \end{pmatrix}. \end{aligned} \tag{4.24}$$

Close to the reference position, the relations between nautical angles, Helmholtz angles, Gibbs vectors, and quaternions can be approximated by the simple formula

$$\begin{pmatrix} \psi \\ \phi \\ \theta \end{pmatrix}_{nautical} \approx \begin{pmatrix} \psi \\ \phi \\ \theta \end{pmatrix}_{Helmholtz} \approx 100 * \begin{pmatrix} r_1 \\ r_y \\ r_3 \end{pmatrix} \approx 100 * \begin{pmatrix} q_1 \\ q_y \\ q_3 \end{pmatrix} \tag{4.25}$$

where θ, ϕ, ψ are given in degrees.

Example

For example, with $\mathbf{r_p} = \begin{pmatrix} 0 \\ 0.176 \\ 0 \end{pmatrix}$ and $\mathbf{r_q} = \begin{pmatrix} 0 \\ 0 \\ 0.087 \end{pmatrix}$, Eq. (4.22) ($\mathbf{r_q} \circ \mathbf{r_p}$) would describe a rotation of 20° about the horizontal axis $\mathbf{s_y}$, followed by a rotation of 10° about the space-fixed vertical axis $\mathbf{s_z}$. According to our discussion above of rotations of objects and coordinate systems, the same formula can also be interpreted as a first rotation of 10° about the vertical axis $\mathbf{b_z}$, followed by a second rotation of 20° about the rotated, object-fixed axis $\mathbf{b_y}$ - which corresponds to the horizontal and vertical rotation in a nautical gimbal.

4.4.2 *Cascaded Rotations with Gibbs Vectors*

For combined rotations, Gibbs vectors show the same sequences as the corresponding rotation matrices or quaternions. Using Gibbs vectors, Eq. (3.39) for combined eye–head movements can be expressed as

$$\mathbf{r}_{gaze} = \mathbf{r}_{head} \circ \mathbf{r}_{eye} \ . \tag{4.26}$$

This can be rearranged to yield the Gibbs vector describing the orientation of our object with respect to the reference frame (e.g. eye in head), $\mathbf{r_{eye}}$, as

$$\mathbf{r}_{eye} = \mathbf{r}_{head}^{-1} \circ \mathbf{r}_{gaze} \ . \tag{4.27}$$

The formula for the multiplication of two Gibbs vectors is given by Eq. (4.22), and the inverse of a Gibbs vector can be determined easily by $\mathbf{r}^{-1} = -\mathbf{r}$.

4.4.3 *Gibbs Vectors and Their Relation to Quaternions*

The Gibbs vector $\mathbf{r}$ which corresponds to the quaternion $\tilde{q}$ describing a rotation of θ about the axis $\mathbf{n}$ is given by

$$\mathbf{r} = \frac{\mathbf{q}}{q_0} = \tan\left(\frac{\theta}{2}\right) \frac{\mathbf{q}}{|\mathbf{q}|} = \tan\left(\frac{\theta}{2}\right) \mathbf{n} \ , \tag{4.28}$$

with $|\mathbf{q}|$ the length of $\mathbf{q}$ as defined in Eq. (4.13).

4.5 Applications

4.5.1 *Targeting an Object in 3-D: Quaternion Approach*

Let us revisit the *aerial gun* application in Sect. 3.6.1, but now assume that we have a targeting device that can be controlled with a quaternion. In other words, the zero quaternion describes the orientation where the targeting device is pointing straight ahead ([1, 0, 0]).

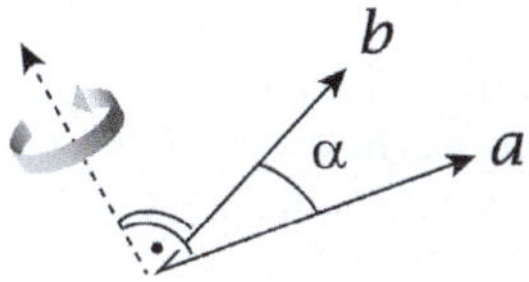

Fig. 4.3 The shortest rotation (α) that brings **a** parallel to **b** is about an axis perpendicular to **a** and **b**

Task: What quaternion would be needed to describe the target orientation, if the target is in an arbitrary location (x, y, z)?

Solution: To answer that question, one can make use of the fact that the shortest rotation that brings a vector **a** into alignment with a vector **b** is a rotation about the direction perpendicular to **a** and **b** (see Fig. 4.3).

$$\mathbf{n} = \frac{\mathbf{a} \times \mathbf{b}}{|\mathbf{a} \times \mathbf{b}|} \tag{4.29}$$

by an angle equal to the angle α between the two vectors

$$\alpha = \arccos\left(\frac{\mathbf{a} \cdot \mathbf{b}}{|\mathbf{a}||\mathbf{b}|}\right) . \tag{4.30}$$

Given the rotation axis and angle, the most convenient way to represent that rotation is the quaternion vector

$$\mathbf{q}_{\text{adjust}} = \mathbf{n} * \sin(\alpha/2) . \tag{4.31}$$

The corresponding algorithm is implemented in `skin.vector.q_shortest_rotation`. For example, if the target moved along an ∞ loop on a screen in 10 m distance, the orientation of the following targeting device could be calculated with the following code:

python **Code**: `C4_targeting.py`: projecting an ∞ loop on a screen (p. 143).

Listing 4.1: C4_targeting.py

```
"""Orientation of targeting device.

"""
# author: Thomas Haslwanter, date: Nov-2017

# Import the required packages
```

```
import numpy as np
import matplotlib.pyplot as plt
import skinematics as skin

# Generate an "infinity"-loop, in 10m distance
t = np.arange(0,20,0.1) # 20 sec, at a rate of 0.1 Hz
y = np.cos(t)
z = np.sin(2*t)
x = 10 * np.ones_like(y)
data = np.column_stack( (x,y,z) )

# Calculate the target-orientation, i.e. the quaternion that
# rotates the vector [1,0,0] such that it points towards
# the target
q_target = skin.vector.q_shortest_rotation([1,0,0], data)

# Plot the results
fig, axs = plt.subplots(2,1)
axs[0].plot(-y,z)
axs[0].set_title('Target on screen, distance=10')
axs[1].plot(q_target)
axs[1].set_xlabel('Time')
axs[1].set_ylabel('Quaternion')
axs[1].legend(['x', 'y', 'z'])
plt.show()
```

4.5.2 Orientation of 3-D Acceleration Sensor

Task: Given an IMU with an accelerometer and a gyroscope only, what is the orientation of the IMU at the beginning of an experiment, based on the direction of gravity indicated by the accelerometer? Specifically, what would be the best guess of the orientation of the sensor in orientation 3 in the example in Fig. 4.4?

Solution: Many inertial sensors are shaped like a match box, and define their long side as the x-axis ($\mathbf{b_x}$), their shorter side as the y-axis ($\mathbf{b_y}$), and the "thick" side as the z-axis ($\mathbf{b_z}$). As explained in more detail in Sect. 2.2.2, a sensor lying flat and stationary on the ground (Fig. 4.4, orientation 1.) indicates an acceleration of

$$\mathbf{acc}_{\text{flat}} = \begin{pmatrix} 0 \\ 0 \\ +9.81 \end{pmatrix} \text{m/s}^2. \tag{4.32}$$

If this sensor is rotated "upright" by exactly 90° (Fig. 4.4, orientation 2.), the readout would indicate (9.81/0/0) m/s^2.

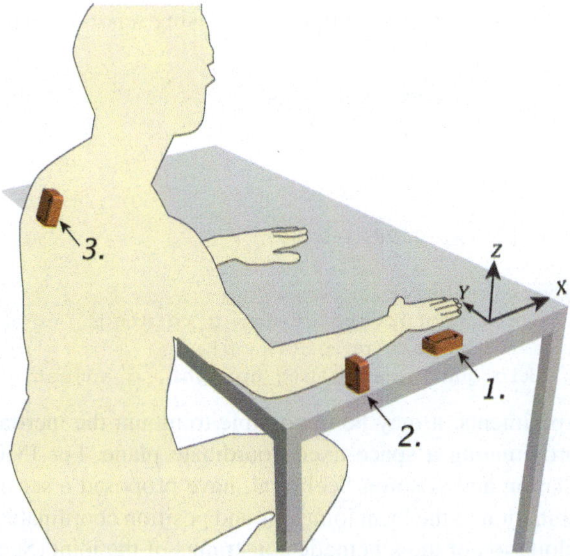

Fig. 4.4 *Orientation 1* Sensor aligned with space-fixed coordinate system. To find the orientation of the sensor on the back, based on the measured accelerations, we first specify the approximate sensor orientation (here *orientation 2*, sensor rotated by 90°). From the measured accelerations in *orientation 3* (sensor on back of subject), the tilt relative to orientation 2 can be determined as described in the text

A rotation "upright" can be indicated by a quaternion vector

$$\mathbf{q}_{\text{upright}} = \begin{pmatrix} 0 \\ -\sin(90°/2) \\ 0 \end{pmatrix} . \tag{4.33}$$

In Fig. 4.4, orientation 3., this sensor is attached in approximately that orientation to the back of an upright standing or sitting person. Since the back of a person is not exactly vertical, the sensor is slightly rotated. What is the best estimate of the orientation of the sensor, when the readout, with the person stationary, indicates e.g. $(9.75/0.98/-0.39)$?

To answer this question, we need the shortest rotation $\tilde{\boldsymbol{q}}_{\text{adjust}}$ that brings the sensor from the "upright" orientation where the accelerometer indicates $(9.81/0/0)$ to the current orientation, where it indicates $(9.75/0.98/-0.39)$. Again, this is the rotation that brings two vectors into alignment, which can be found as in the example above.

Since a rotation about $\mathbf{g}$ does not change the output of the accelerometer, the best estimate of the orientation of the accelerometer is

$$\tilde{\boldsymbol{q}}_{\text{total}} = \tilde{\boldsymbol{q}}_{\text{upright}} \circ \tilde{\boldsymbol{q}}_{\text{adjust}} , \tag{4.34}$$

where $\circ$ indicates the quaternion multiplication. Using *scikit-kinematics*, this can be implemented as

```
# Import the required packages
import skinematics as skin

# Enter the measurements
g = [9.81, 0, 0]
g_upright = [9.75, 0.98, -0.39]

# Calculate the sensor orientation
q_adjust = skin.vector.qrotate(g, g_rotated)
q_upright = [0, np.sin(np.pi/4), 0]
q_total = skin.quat.quatmult(q_upright, q_adjust)
```

For some experiments, it may be impossible to mount the inertial sensors in an orientation approximating a space-fixed coordinate plane. For IMUs in arbitrary mounting orientation and position Seel et al. have proposed a set of methods that allow the determination of the local joint axis and position coordinates from arbitrary motions by exploitation of the kinematic constraints of the joint (Seel et al. 2014).

4.5.3 Calculating Orientation of a Camera on a Moving Object

Consider the problem where a camera in a missile must be pointed to look at a specific target. The missile attitude/orientation has three rotational degrees of freedom relative to the world. The camera attitude/orientation also has three rotational degrees of freedom but relative to the missile body. So the missile and camera gimbal forms a set of cascaded three-axis transformations. The camera gaze direction is $(1, 0, 0)$, i.e., the optical axis is along the x direction. When the camera is looking at the object in the world, the target's location in camera coordinates must, therefore, be $(x_{\text{obj}}^{\text{camera}}, 0, 0)$.

There are three coordinate systems in this scenario: (1) fixed to the world, (2) fixed to the missile body, and (3) fixed to the camera on the gimbal. The target direction in world coordinates is known from the location of the missile and the target. The target vector in the camera coordinates is $\mathbf{t}^c = [|\mathbf{t}|, 0, 0]$, i.e., the optical axis or gaze direction. The target direction in the missile body coordinates can be calculated.

In the example below, a missile is located at position (10, 1700, −2200) m with an attitude (pitch = −1.2, yaw = −0.2, roll = −1.1) rad in Helmholtz sequence (roll–yaw–pitch from outside to inside). The target is located at position (23, −560, −1800) m. How can one calculate the Helmholtz-sequence camera gimbal attitude, relative to the missile body, such that the camera optical axis points at the target? (To be on the optical axis in camera coordinates, the target vector in camera coordinates must be (2295, 0, 0).)

4.5.3.1 Calculating Look-at Angles

python **Code**: `C4_look_at.py`: How to calculate the orientation of a camera on a missile, in order to look in the direction of a given target.

Listing 4.2: C4_look_at.py

```
""" Given the positions of a missile and a target, and the
missile orientation, calculate the gimbal orientation of a
camera mounted on the missile, such that the camera
points at the target.
The optical axis of the camera is the x-axis.
"""

# author: ThH, date: July, 2018, ver:  1.1

# Import the required packages
import numpy as np
import skinematics as skin


def camera_orientation(missile_pos, missile_orient,
    target_pos):
    '''Find camera orientation re missile, to focus on target.

    Inputs
    ------
    missile_pos : ndarray (3,) or (N,3)
        Position of missile in space
    missile_orient : ndarray (3,) or (N,3)
        Orientation of missile, in Helmholtz angles [rad]
    target_pos : ndarray (3,) or (N,3)
        Position of target in space

    Returns
    -------
    camera_orientation : ndarray (3,) or (N,3)
        Camera orientation, in Helmholtz angles [deg]
    '''

    # Required camera direction in space is a vector from
        missile to target
    v_missile_target = target_pos - missile_pos

    # Camera direction re missile
    q = skin.rotmat.seq2quat(np.rad2deg(missile_orient),
    seq='Helmholtz')
    tm_in_missile_CS = skin.vector.rotate_vector
    (v_missile_target, -q)
```

```
    # Required camera orientation on missile, to focus on the
        target
    camera_orientation = skin.vector.target2orient
    (tm_in_missile_CS, orient_type='Helmholtz')

    return camera_orientation


if __name__=='__main__':

    # Set up the system
    helm = [-1.2, -0.2, -1.1] # Missile orientation, in
        Helmholtz angles [rad]
    target = np.r_[10, 1700, -2200]
    missile = np.r_[23, -560, -1800]

    # Find the camera orientation
    camera = camera_orientation(missile, helm, target)

    # Show the results
    print('Camera orientation on missile, in Helmholtz
          angles:\n' +
          'pitch={0:4.2f}, yaw={1:4.2f} [deg]'.
           format(*camera))
```

4.5.4 *Object-Oriented Implementation of Quaternions*

The Python module `scikit.quat` also contains a class `Quaternion` with multiplication, division, and inversion. A Quaternion can be created from vectors, rotation matrices, or from nautical angles, Helmholtz angles, or Euler angles. It provides operator overloading for `mult`, `div`, and `inv`, indexing, and access to the data in the attribute `values`.

```
import numpy as np
from skinematics.quat import Quaternion

data = np.array([[0,0,0.1], [0, 0.2, 0]])
data2 = np.array([[0,0,0.1], [0,0,0.1]])

eye = Quaternion(data)
head = Quaternion(data2)
gaze = head * eye
print(gaze)
#>> Quaternion [[ 0.98        0.         0.          0.19899749]
#>>             [ 0.97488461 -0.02       0.19899749 0.09797959]]
```

python Code: `C4_examples_quat.py`: Examples of working with quaternions: quaternion multiplication, conjugation, inversion, etc. (p. 144)

Chapter 5
Velocities in 3-D Space

Typical biomechanical problems involve not only knowledge about the orientation of an object, but also knowledge about angular and linear position, velocity, and acceleration. Consider the movement of a rigid object in space, for example, the lower arm of a person. How can the forces to obtain the observed movement in space be determined?

The equations of motion involve linear and angular acceleration. This chapter shows how these can be obtained from position and orientation, and vice versa, how position and orientation can be calculated from linear and angular velocity.

5.1 Equations of Motion

The movement of an object must fulfill the following equations:

1. For translations of the center-of-mass (COM) of an object with mass m, *Newton's second law*

$$\sum_i \mathbf{f}_i = \frac{d(m * \mathbf{vel})}{dt} = m * \mathbf{acc} \tag{5.1}$$

   where $\mathbf{f}_i$ are the forces acting on the object, and **vel** and **acc** the velocity and acceleration of the COM, respectively.
2. For rotations about the center of mass, with constant moments of inertia

$$\sum \mathbf{m}_i = \frac{d(\mathbf{\Theta} \cdot \boldsymbol{\omega})}{dt} = \mathbf{\Theta} \cdot \frac{\mathrm{d}\boldsymbol{\omega}}{\mathrm{d}t}. \tag{5.2}$$

where $\boldsymbol{\omega}$ is the angular velocity vector, and $\mathbf{m}_i$ are the moments acting on the body.

T. Haslwanter, *3D Kinematics*, https://doi.org/10.1007/978-3-319-75277-8_5

For example, for limb movements only two types of forces are acting on the limb: mass forces, i.e., gravity $m * \mathbf{g}$; and the proximal and distal *joint reaction forces* $\mathbf{r}_{\text{prox}}$ and $\mathbf{r}_{\text{dist}}$, i.e., the forces acting on the joints.

$$\mathbf{r}_{\mathbf{prox}} + \mathbf{r}_{\mathbf{dist}} + m * \mathbf{g} = m * \mathbf{acc}. \tag{5.3}$$

Once the linear and angular accelerations of the object, **acc** and $\frac{d\omega}{dt}$, are known, Eqs. 5.1 and 5.2 can be used to calculate the forces and the moments involved, and thus also the work required to move the object in the observed way.

This chapter will show how **acc** and $\frac{d\omega}{dt}$ can be found. In order to calculate them from measurement signals, the equations from the previous chapters will be required. Specific tracking systems, e.g. optical or inertial based systems, will be dealt with in the next chapter.

5.2 Linear Velocity

The mathematical description of translations is fairly simple, since translations are always commutative. In other words, one always arrives at the same position, regardless of the sequence in which the translations are executed (Fig. 5.1):

$$\mathbf{pos_x} + \mathbf{pos_y} = \mathbf{pos_y} + \mathbf{pos_x}. \tag{5.4}$$

This has important consequences. One is that velocity and acceleration for the translational movement components can be calculated for each axis separately:

$$\begin{aligned} &\mathbf{vel} = \frac{d\mathbf{pos}}{dt} \\ &\mathbf{acc} = \frac{d^2\mathbf{pos}}{dt^2} \\ &\text{with } vel_i = \frac{dpos_i}{dt}, \; and \quad acc_i = \frac{d^2pos_i}{dt^2} \; (i = 1, 2, 3). \end{aligned} \tag{5.5}$$

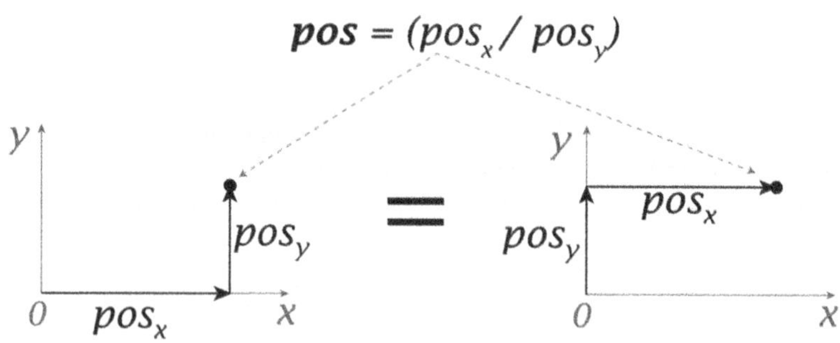

Fig. 5.1 Translations are commutative

For example, one can calculate when a bullet shot from a gun will fall down on the ground at $-z$, without knowing how fast it is flying forward along x.

This will not be the case for rotations!

Depending on the computational requirements, a number of methods are available to implement Eq. (5.5). For example, the *Savitzky–Golay Filter*, which is available in Python and in Matlab,[1] offers a very efficient and convenient way to smooth and to differentiate data, especially for interactive work with experimental data. If the `pos` data and the sample rate are given, velocity and acceleration can be calculated with

```
from scipy.signal import savgol_filter
dt = 1/rate
vel = savgol_filter(pos, window_length=15, polyorder=3,
deriv=1,delta=dt,axis=0)
acc = savgol_filter(pos, window_length=15, polyorder=3,
deriv=2,delta=dt,axis=0)
```

The parameter `window_length` controls the amount of smoothing, and the last argument, `axis=0`, is only required if data are stored in column-form.

Savitzky–Golay filters are similar to "finite difference approximations", which fit polynomials *exactly* to a given data window, and calculate the derivatives of these polynomials exactly. This leads to tables of "finite difference coefficients over n-point stencils", such as the ones listed in (Diebel 2006). But in contrast to those exact calculations, Savitzky–Golay filters determine the *best-fit* polynomials, typically to larger sample windows, and can also calculate the derivatives from those best-fit polynomials. Surprisingly, this can be done with a fixed Finite Impulse Response (FIR) filter, whose coefficients only depend on the window size, the order of the polynomial fit, the order of the derivative, and the sampling rate. This leads to a very fast and efficient implementation of this filter.

In practice the chosen `window_length` is often checked visually, to ensure that there is, on the one hand, sufficient smoothing to eliminate the noise, and on the other hand that the important signal characteristics are still kept.

Also note that for velocities and accelerations, the reference position no longer makes any difference, since the differentiation in Eq. 5.5 eliminates any offset.

Determination of Position from Acceleration

To understand how angular velocity can be converted into orientation in space, remember first how linear acceleration can be converted into linear velocity, and linear velocity in turn into position.

[1] An easy-to-use Matlab implementation for the Savitzky–Golay filter for smoothing and derivatives is the command `savgol` in the Matlab *Kinematics Toolbox*.

$$\mathbf{vel}(t) = \mathbf{vel}(t_0) + \int_{t_0}^{t} \mathbf{acc}(t')\,dt' \tag{5.6}$$

$$\mathbf{x}(t) = \mathbf{x}(t_0) + \int_{t_0}^{t} \mathbf{vel}(t'')\,dt'' =$$

$$= \mathbf{x}(t_0) + \mathbf{vel}(t_0) * (t - t_0) + \int_{t_0}^{t}\int_{t_0}^{t''} \mathbf{acc}^{\text{space}}(t')\,dt'\,dt''. \tag{5.7}$$

Starting with a stationary sensor, i.e., $\mathbf{vel}(t_0) = 0$, the change in position is given by

$$\Delta\mathbf{x}(t) = \mathbf{x}(t) - \mathbf{x}(t_0) = \int_{t_0}^{t}\int_{t_0}^{t''} \mathbf{acc}^{\text{space}}(t')\,dt'\,dt''. \tag{5.8}$$

When working with computers, the integral cannot be performed exactly, but only approximately. Employing the Riemann–Stieltjes integral theorem, splitting the time between t_0 and t into n equal elements with width Δt leads to

$$\mathbf{x}(t) = \mathbf{x_0} + \Delta\mathbf{x_1} + \Delta\mathbf{x_2} + \cdots + \Delta\mathbf{x_n}. \tag{5.9}$$

Measuring the acceleration at discrete times $t_i (i = 0, .., n)$, Eqs. (5.6) and (5.7) have to be replaced with discrete equations:

$$\mathbf{vel}(t_{i+1}) \approx \mathbf{vel}(t_i) + \mathbf{acc}^{\text{space}}(t_i) * \Delta t \tag{5.10}$$

$$\mathbf{x}(t_{i+1}) \approx \mathbf{x}(t_i) + \mathbf{vel}(t_i) * \Delta t + \frac{\mathbf{acc}^{\text{space}}(t_i)}{2} * \Delta t^2 \tag{5.11}$$

with the sampling period Δt. The solution to Eq. (5.9) with the approximation in Eq. (5.11) can be conveniently obtained in Matlab and Python with the `cumtrapz` command, which Python users can find in the module `scipy.integrate`. More stability can be achieved by modifying Eq. (5.10) with the Euler–Cromer method (see Exercise 1.3).

Here it is important to note that due to the commutativity of translations, the sequence of additions has no effect.

$$\mathbf{x_0} + \Delta\mathbf{x_1} + \Delta\mathbf{x_2} + \cdots + \Delta\mathbf{x_n} = \Delta\mathbf{x_n} + \Delta\mathbf{x_{n-1}} + \cdots + \Delta\mathbf{x_1} + \mathbf{x_0}\,. \tag{5.12}$$

5.3 Angular Velocity

5.3.1 Calculating Angular Velocity from Orientation

Just as the value of the current position depends on the chosen reference position, 3-D orientation depends on the choice of the reference orientation. In contrast, the inertial *angular velocity* does not depend on the reference orientation, since it only describes the movement from the current orientation to the next, which does not involve the reference orientation.

Quaternion Angular Velocity

The simplest formula describing the angular velocity $\boldsymbol{\omega}$ is given in the quaternion notation:

$$\tilde{\omega} = 2 * \frac{\mathrm{d}\tilde{\boldsymbol{q}}}{\mathrm{d}t} \circ \tilde{\boldsymbol{q}}^{-1}, \tag{5.13}$$

where $\tilde{\omega} = (0, \boldsymbol{\omega})$ is a (*pure*) quaternion, with the vector $\boldsymbol{\omega}$ describing the 3-D angular velocity with respect to space. Note that due to the noncommutativity of rotations the angular velocity depends not only on the time-derivative $\frac{\mathrm{d}\tilde{\boldsymbol{q}}}{\mathrm{d}t}$ of the orientation, but also on the current orientation $\tilde{\boldsymbol{q}}$.

Angular Velocity Tensor

If the orientation is described with rotation matrices, the calculation of angular velocity is less straightforward. Any vector $\mathbf{p}$ that rotates around an axis with angular velocity $\boldsymbol{\omega}$ satisfies:

$$\frac{\mathrm{d}\mathbf{p}(t)}{\mathrm{d}t} = \boldsymbol{\omega} \times \mathbf{p}\,. \tag{5.14}$$

The cross-product, which is a nonlinear operation, can be eliminated by describing angular velocity with a matrix, the so-called "angular velocity tensor" $\boldsymbol{\Omega}$. The components of the angular velocity $\boldsymbol{\omega}$ correspond to the off-diagonal elements of the angular velocity tensor as follows:

$$\boldsymbol{\Omega}(t) = \begin{bmatrix} 0 & -\omega_z(t) & \omega_y(t) \\ \omega_z(t) & 0 & -\omega_x(t) \\ -\omega_y(t) & \omega_x(t) & 0 \end{bmatrix}. \tag{5.15}$$

This tensor $\boldsymbol{\Omega}(t)$ will act as if it were a $(\boldsymbol{\omega}\times)$ operator :

$$\boldsymbol{\omega}(t) \times \mathbf{p}(t) = \boldsymbol{\Omega}(t)\mathbf{p}(t)\,. \tag{5.16}$$

Given the orientation matrix $\mathbf{R}$(t) of a frame, its instant angular velocity tensor $\boldsymbol{\Omega}$ (t) can be obtained as follows. We know that

$$\frac{\mathrm{d}\mathbf{p}(t)}{\mathrm{d}t} = \boldsymbol{\Omega}(t) \cdot \mathbf{p}\,. \tag{5.17}$$

As the angular velocity for the three basis vectors of a rotating frame must be the same, if we have a matrix **R**(t) whose columns are the vectors of the frame, we can write for the three vectors as a whole:

$$\frac{\mathrm{d}\mathbf{R}(t)}{\mathrm{d}t} = \boldsymbol{\Omega} \cdot \mathbf{R}(t)\,. \tag{5.18}$$

As a result, the angular velocity tensor is

$$\boldsymbol{\Omega} = \frac{\mathrm{d}\mathbf{R}(t)}{\mathrm{d}t} \cdot \mathbf{R}^{-1}(t)\,. \tag{5.19}$$

Note the similarity between Eqs. (5.13) and (5.19)!

From Gibbs Vectors

Expressed in Gibbs vectors (Hepp 1990), Eq. (5.13) is equivalent to

$$\boldsymbol{\omega} = 2 * \frac{\frac{\mathrm{d}\mathbf{r}}{\mathrm{d}t} + \mathbf{r} \times \frac{\mathrm{d}\mathbf{r}}{\mathrm{d}t}}{1 + \mathbf{r}^2}\,. \tag{5.20}$$

From Nautical Angles

A more complex formula is required if angular velocity is expressed in nautical angles (Goldstein 1980):

$$\boldsymbol{\omega} = \begin{pmatrix} \frac{\mathrm{d}\psi_N}{\mathrm{d}t} * \cos\theta_N * \cos\varphi_N & - \frac{\mathrm{d}\varphi_N}{\mathrm{d}t} * \sin\theta_N \\ \frac{\mathrm{d}\varphi_N}{\mathrm{d}t} * \cos\theta_N & + \frac{\mathrm{d}\psi_N}{\mathrm{d}t} * \sin\theta_N * \cos\varphi_N \\ \frac{\mathrm{d}\theta_N}{\mathrm{d}t} & - \frac{\mathrm{d}\psi_N}{\mathrm{d}t} * \sin\varphi_N \end{pmatrix} \tag{5.21}$$

Equations (5.13) and (5.21) are equivalent, as they express the same angular velocity with different parametrizations. The time-derivatives of the orientation coordinates – $\frac{d}{dt}\tilde{\boldsymbol{q}}$ for quaternions, $\frac{d}{dt}\mathbf{r}$ for Gibbs vectors, and

$$\begin{pmatrix} \frac{\mathrm{d}\theta_N}{\mathrm{d}t} \\ \frac{\mathrm{d}\phi_N}{\mathrm{d}t} \\ \frac{\mathrm{d}\psi_N}{\mathrm{d}t} \end{pmatrix} = \frac{\mathrm{d}}{dt} \begin{pmatrix} \theta \\ \phi \\ \psi \end{pmatrix}_N$$

for nautical angles—are often referred to as “coordinate velocities”. The coordinate velocity obviously depends on the parameters chosen to describe the orientation. In contrast, the inertial angular velocity vector $\boldsymbol{\omega}$ describes the actual movement of the object, with its axis given by the instantaneous axis of rotation, and its length by the angular velocity of this rotation, and does not depend on the parametrization of the orientation. The preceding formulas also show that the coordinate velocity is in general not equivalent to the inertial angular velocity $\boldsymbol{\omega}$.

Relation Angular Velocity-Orientation

The noncommutativity of rotations can lead to seemingly counterintuitive phenomena. One such phenomenon arises in oculomotor research. If quaternion vectors are used to describe the 3-D orientation of the eye, one finds empirically that under typical conditions (i.e., looking around, with the head upright and stationary) all quaternion vectors describing eye orientations line up along a plane. This plane is called "Listing's plane" (Haslwanter 1995). But if the eye rotates from one eye orientation (in Listing's plane) to another eye orientation (also in Listing's plane), it has to rotate about an axis that, in general, does not lie in that plane! This is illustrated graphically in Fig. 5.2, and is a consequence of the fact that the quaternion multiplication required to calculate angular velocity (Eq. 5.13) involves a cross-product of the quaternion vectors and their time-derivative (Eq. 4.6). Since the tilt of the eye velocity vector in the example indicated in Fig. 5.2 is half the eccentricity of the eye orientation, this effect is in oculomotor research referred to as "half-angle rule".

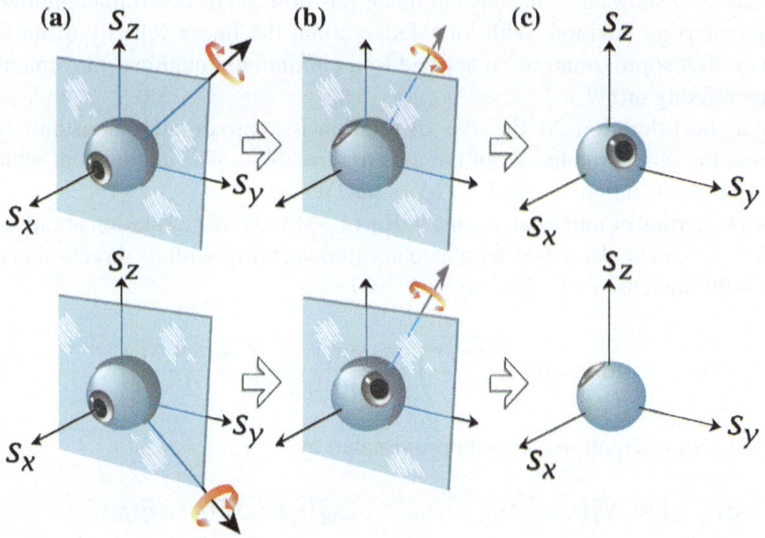

Fig. 5.2 **a** Eye in the reference orientation. Rotations from **a** to **b**: Under typical conditions, quaternion vectors describing eye orientation lie in a plane called "Listing's plane". Here this plane is perpendicular to gaze in the reference orientation. Therefore, eye positions in (**b**) have been reached from the reference position in (**a**) by a rotation about axes that lie in that plane, indicated with solid arrows in (**a**). But to get from one eccentric eye orientation to another, i.e., **b** → **c**, the eye has to rotate about an axis which does not lie in that plane

For example, the quaternion vectors of the unit quaternions for $\tilde{q}_{start}$ looking approximately 20° up and 20° right, and for $\tilde{q}_{end}$ looking approximately 20° up and 20° left are

$$\mathbf{q}_{start} = \begin{pmatrix} 0 \\ -0.2 \\ -0.2 \end{pmatrix}, \; \mathbf{q}_{end} = \begin{pmatrix} 0 \\ -0.2 \\ 0.2 \end{pmatrix}. \tag{5.22}$$

Both have a 0 component for q_1. But the rotation from $\tilde{q}_{start}$ to $\tilde{q}_{end}$ is given by

$$\tilde{q}_{move} = \tilde{q}_{end} \circ \tilde{q}_{start}^{-1} = \begin{pmatrix} 0.92 \\ -0.08 \\ 0 \\ 0.38 \end{pmatrix}. \tag{5.23}$$

Since the axis of this quaternion is also the axis of eye velocity, the eye velocity is tilting backward, despite the fact that neither $\mathbf{q}_{start}$ nor $\mathbf{q}_{end}$ are.

5.3.2 Calculating Orientation from Angular Velocity

Equation (5.9) shows how the current linear position can be determined numerically from the starting position, with knowledge about the linear velocity of an object. How can this approximation be applied to a continuous rotational movement with angular velocity $\boldsymbol{\omega}(t)$?

For a short duration Δt the axis of rotation is approximately constant. During this time, the object rotates about the axis $\mathbf{n}(t) = \frac{\boldsymbol{\omega}(t)}{|\boldsymbol{\omega}(t)|}$. The angle about which the object rotates during this period is $\Delta\phi = |\boldsymbol{\omega}(t)| \cdot \Delta t$.

The properties of unit quaternions in Eq. (4.13) state that a rotation about $\mathbf{n}$ by an angle of $\Delta\phi$ can be described with a quaternion vector $\mathbf{q}$, with its direction given by $\mathbf{n}$, and with the length $\sin(\frac{\Delta\phi}{2})$. So with

$$\Delta\mathbf{q}(t_i) = \mathbf{n}(t_i) \sin\left(\frac{\Delta\phi(t_i)}{2}\right) = \frac{\boldsymbol{\omega}(t_i)}{|\boldsymbol{\omega}(t_i)|} \sin\left(\frac{|\boldsymbol{\omega}(t_i)|\,\Delta t}{2}\right), \tag{5.24}$$

the current orientation can be approximated by

$$\tilde{q}(t_{n+1}) \approx \Delta\tilde{q}(t_n) \circ \Delta\tilde{q}(t_{n-1}) \circ \ldots \circ \Delta\tilde{q}(t_2) \circ \Delta\tilde{q}(t_1) \circ \tilde{q}(t_0). \tag{5.25}$$

Note that Eq. (5.25) contains full quaternions (written as $\tilde{q}$), while Eq. (5.24) defines only the corresponding quaternion vectors (written as $\mathbf{q}$).

Important Note: In Eq. (5.25) the sequence is important, because rotations are noncommutative! Here $\boldsymbol{\omega}$ is the angular velocity *with respect to inertial space*. In the next chapter it will be shown how Eq. (5.25) has to be modified if $\boldsymbol{\omega}$ is the angular velocity as seen from a rotating body.

python™ **Code**: `C5_examples_vel.py`: Worked example of how to calculate angular velocity from 3-D orientation and vice versa (p. 147).

Chapter 6
Analysis of 3-D Movement Recordings

The first section of this chapter 6.1 will investigate how movement parameters can be determined with marker-based systems. Optical Recording Systems (ORS), like the system from Zeiss shown in Fig. 6.1, can provide high-resolution position information for markers attached to objects. This allows to directly determine object position and orientation in space. In contrast, Inertial Measurement Units (IMUs), which will be described in more detail in the next Sect. 6.2, indicate linear acceleration and angular velocity and contain no direct information about the absolute position and orientation of the object in space.

6.1 Position and Orientation from Optical Sensors

A good overview of the conceptual background underlying the reconstruction of human skeletal motion is given in the *Handbook of Human Motion*, in the chapter by (Camomilla and Vannozzi 2018). And the estimation of dynamic 3-D pose based on optical motion capture systems is described in (Selbie and Brown 2018). The presentation here focusses on the kinematic principles underlying the 3-D analysis.

6.1.1 Recording 3-D Markers

To define position and orientation of an object in three dimensions, one needs to find the positions of three points $\mathbf{p_i}(t)$ that are firmly connected to the object. The only requirements for these points are that they (a) are visible, and (b) must not be arranged along a line.

In the example sketched out in Fig. 6.2, three markers are attached to the left lower arm. Assume that the positions of these markers have been recorded, and stored as

T. Haslwanter, *3D Kinematics*, https://doi.org/10.1007/978-3-319-75277-8_6

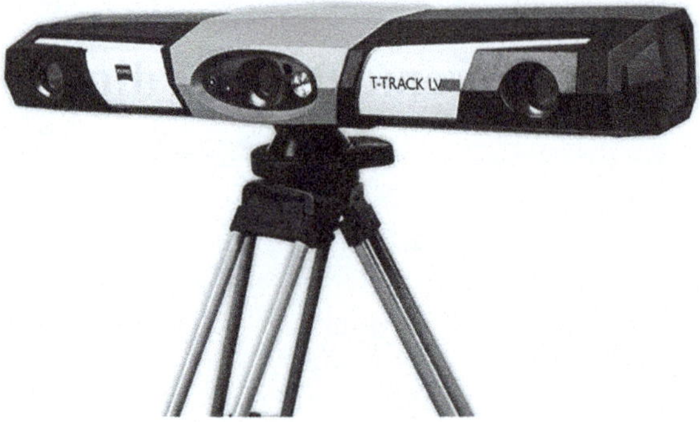

Fig. 6.1 Optical recording system (with kind permission from *Carl Zeiss Optotechnik GmbH*)

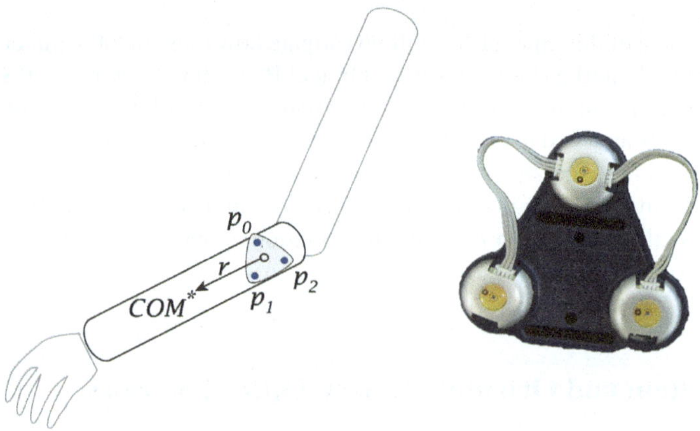

Fig. 6.2 Recording movements of the lower arm with optical markers. (Left) $\mathbf{p}_i$ indicate the position of the markers, "o" the middle of the markers, and "*" the location of the Center of Mass (COM). (Right) Active markers for 3-D position measurements, for the *Optotrak*-system

$\mathbf{p_i}(t), i = 0, 1, 2$. To investigate the object dynamics, the following questions have to be answered:

1. What are the positions $\mathbf{x}(t)$, linear velocities $\mathbf{v}(t) = \frac{\mathrm{d}\mathbf{x}(t)}{\mathrm{d}t}$, and linear accelerations $\mathbf{acc}(t) = \frac{\mathrm{d}^2\mathbf{x}(t)}{\mathrm{d}t^2}$ of the markers, with respect to our chosen space-fixed coordinate system?
2. What are the resulting orientation $\mathbf{R}(t)$, angular velocity $\boldsymbol{\omega}(t)$, and angular acceleration $\frac{\mathrm{d}\boldsymbol{\omega}}{\mathrm{d}t}$ of the markers?
3. What are the locations of the markers relative to the point(s) of interest on the object, in an object-fixed reference system?

The *position* of an object is typically taken to be its "center of mass" (COM), sometimes also called "center of gravity". The COM is in general given by

$$\mathbf{COM}(t) = \frac{\sum_{i=0}^{n-1} m_i * \mathbf{x}_i(t)}{\sum_{i=0}^{n-1} m_i} \tag{6.1}$$

where $\mathbf{x}_i$ are the locations of the mass elements m_i. Since the three markers have the same "weight" m_i, the position of the center of the markers can be calculated as

$$\mathbf{m}(t) = \frac{\sum_{i=0,1,2} \mathbf{p}_i(t)}{3} \tag{6.2}$$

In Fig. 6.2, the position of the center of the markers is indicated with "o". With $\mathbf{c}(t)$ defined as the location of the COM, the vector $\mathbf{r}(t)$ from the markers to the COM is given by

$$\mathbf{r}(t) = \mathbf{c}(t) - \mathbf{m}(t) \tag{6.3}$$

6.1.2 Orientation in Space

To find the orientation of an object, one needs to find the rotation matrix **R** describing the orientation of the object. Since a rotation matrix is given by three columns of orthogonal unit vectors, one needs to find three orthogonal unit vectors which are uniquely defined through the marker points $\mathbf{p_i}(t)$ (see Fig. 6.3).

Let the center of the local, marker-fixed—and thereby object-fixed—coordinate system be determined by $\mathbf{p_0}$, and the direction of the positive first coordinate axis by the vector $\overrightarrow{\mathbf{p_0 p_1}}$. In general, the line $\overrightarrow{\mathbf{p_0 p_2}}$ is not perpendicular to $\overrightarrow{\mathbf{p_0 p_1}}$. So in order to uniquely define a normalized, right-handed coordinate system characterizing position and orientation of an object, one can use a procedure called "Gram–Schmidt Orthogonalization"[1]:

$$\begin{aligned} \mathbf{a_x} &= \frac{\mathbf{p_1}-\mathbf{p_0}}{|\mathbf{p_1}-\mathbf{p_0}|} \\ \mathbf{a_z} &= \frac{\mathbf{a_x}\times(\mathbf{p_2}-\mathbf{p_0})}{|\mathbf{a_x}\times(\mathbf{p_2}-\mathbf{p_0})|} \\ \mathbf{a_y} &= \mathbf{a_z} \times \mathbf{a_x}. \end{aligned} \tag{6.4}$$

The three orthogonal unit vectors $\mathbf{a_i}(t)$ define the columns of the rotation matrix **R** which describes the orientation of the object

$$\mathbf{R}(t) = \left[\mathbf{a_x}(t)\ \mathbf{a_y}(t)\ \mathbf{a_z}(t) \right]. \tag{6.5}$$

[1]An alternative way to perform a Gram–Schmidt orthogonalization is given in Appendix A.2.

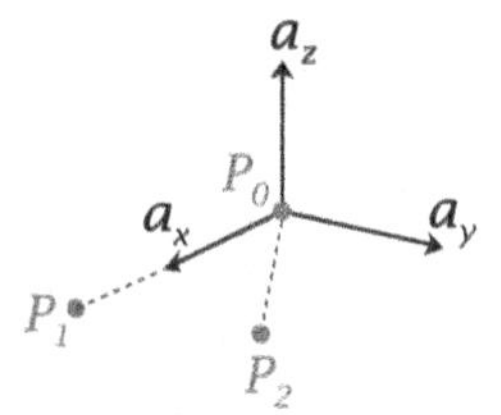

Fig. 6.3 Gram–Schmidt Orthogonalization: calculation of three orthogonal unit vectors $[\mathbf{a}_x, \mathbf{a}_y, \mathbf{a}_z]$, uniquely defined by three points $[\mathbf{p}_0, \mathbf{p}_1, \mathbf{p}_2]$

Denoting the rotation of the object relative to the starting (or reference) orientation at $t = t_0$ with $\mathbf{R}_{mov}$ leads to

$$\mathbf{R}(t) = \mathbf{R}_{mov}(t) \cdot \mathbf{R}(t_0). \tag{6.6}$$

Bringing $\mathbf{R}(t_0)$ to the other side of Eq. (6.6), the rotation matrix characterizing the rotational movement is

$$\mathbf{R}_{mov}(t) = \mathbf{R}(t) \cdot \mathbf{R}(t_0)^{-1}. \tag{6.7}$$

Note that rotation matrices are not the only way to describe the orientation of an object, and/or its angular movement. The same orientation/movement can also be described with equivalent quaternions (Eq. 4.14) or Gibbs vectors (Eq. 4.20).

6.1.3 Position in Space

Once the location of the markers and their orientation is known, the position of the COM (or any other point of interest) of the object is uniquely defined, even if the markers are mounted eccentrically to the COM. The movement of every point is given by the sum of the COM translation *plus* its rotation about the COM. With

$$\mathbf{r}(t) = \mathbf{R}_{mov}(t) \cdot \mathbf{r}(t_0) \tag{6.8}$$

and using Eq. (6.3), the movement of the COM is given by

$$\mathbf{c}(t) = \mathbf{m}(t) + \mathbf{r}(t) = \mathbf{m}(t) + \mathbf{R}_{mov}(t) \cdot \mathbf{r}(t_0). \tag{6.9}$$

This finally provides the 3-D position of the point of interest on the object.

6.1.4 *Velocity and Acceleration*

Linear Velocity and Acceleration
The equations for finding linear velocity and linear acceleration of an object are simple. If the position of an object is denoted with $\mathbf{pos}(t)$, linear velocity and acceleration are given by (see Sect. 5.2):

$$\mathbf{vel}(t) = \frac{\mathrm{d}\mathbf{pos}(t)}{\mathrm{d}t} \qquad (6.10)$$
$$\mathbf{acc}(t) = \frac{\mathrm{d}^2\mathbf{pos}(t)}{\mathrm{d}t^2}.$$

Angular Velocity
From the orientation of an object the corresponding angular velocity $\boldsymbol{\omega}$ can be calculated. Expressing the orientation with a quaternion $\tilde{\boldsymbol{q}}$, the angular velocity $\boldsymbol{\omega}$ can be found with Eq. (5.13). Equivalently, when the orientation is described with Gibbs vectors, the angular velocity can be found with Eq. (5.20); and expressing the orientation with the nautical sequence, the angular velocity can be found with Eq. (5.21).

Angular Acceleration
The angular acceleration can be obtained from the angular velocity through simple differentiation

$$\mathbf{AngAcc} = \frac{\mathrm{d}\boldsymbol{\omega}}{\mathrm{d}t}. \qquad (6.11)$$

Note that while the noncommutativity of rotations requires more complex formulas to find the *angular velocity* from orientation, the *angular acceleration* is simply the time derivative of angular velocity!

6.1.5 *Transformation from Camera- to Space-Coordinates*

Orientation
The first step in the analysis of 3-D movement recordings is the determination of the position and orientation of the ORS with respect to the chosen space-fixed coordinate system.

The output of the data recorded by the ORS cannot be used directly, because they are relative to the recording system. Commonly the (**x**/**y**) directions for the recorded data are determined by the image plane of the optical sensor. **x** indicates the horizontal image direction, **y** the vertical image direction, and **z** completes a right-handed coordinate system $CS_{\mathrm{ORS}} = [\mathbf{x}\,\mathbf{y}\,\mathbf{z}]$ (see Fig. 6.4).

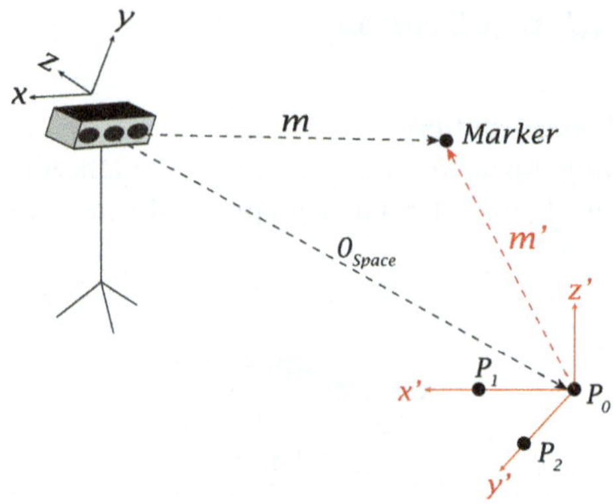

Fig. 6.4 Position of a *Marker* as seen from the ORS (**m**), and as seen from a space-fixed coordinate system (**m**′)

To identify the position and orientation of the ORS relative to a space-fixed coordinate system $CS_{\text{space}} = [\mathbf{x}'\ \mathbf{y}'\ \mathbf{z}']$, the following procedure can be applied: Let three marker points be positioned along the space-fixed coordinate system CS_{space}, such that

1. $\mathbf{p_0}$ is at the center of CS_{space}.
2. The vector from $\mathbf{p_0}$ to $\mathbf{p_1}$ defines the space-fixed x-axis, $\mathbf{x}'$.
3. The plane defined by $(\mathbf{p}_0, \mathbf{p}_1, \mathbf{p}_2)$ defines the space-fixed *x/y*-plane. (In practice, it is helpful if the vector from $\mathbf{p_0}$ to $\mathbf{p_2}$ points approximately in the direction of $\mathbf{y}'$.)

$\mathbf{p_0}$ indicates the position of the origin of CS_{space} with respect to the ORS. Using a Gram–Schmidt orthogonalization, the rotation matrix **R** indicating the orientation of CS_{space} with respect to CS_{ORS} is given by (see Fig. 6.4):

$$\mathbf{R} = [\mathbf{x}'\ \mathbf{y}'\ \mathbf{z}']. \tag{6.12}$$

6.1.6 *Position*

Relative to ORS: If the position of the marker as seen from the ORS is **m**, and the location of the center of CS_{space} is indicated by $\mathbf{p_0}$, then the vector from the center of CS_{space} to the marker **m** is given by

$$\mathbf{m}'_{ORS} = -\mathbf{p_0} + \mathbf{m}. \tag{6.13}$$

Note that at this point the components of the vector are still expressed in the orientation of CS_{ORS}!!

Relative to Space: To get the components of $\mathbf{m}'$ relative to CS_{space}, one has to take the orientation of CS_{space} with respect to CS_{ORS} into consideration, to obtain

$$\mathbf{m}'_{\text{space}} = \mathbf{R}^{-1} * \mathbf{m}'_{\text{ORS}}. \tag{6.14}$$

Combining Eqs. (6.13) and (6.14) gives

$$\mathbf{m}'_{\text{space}} = \mathbf{R}^{-1} * (\mathbf{m} - \mathbf{p_0}). \tag{6.15}$$

6.2 Position and Orientation from Inertial Sensors

This section describes the calculation of the exact orientation and position in space, given IMU data from ideal linear accelerometers and gyroscopes (Sect. 6.2.1). The next chapter ("Sensor Integration", Chap. 7) will present how "Kalman Filters" or other analysis procedures can be used to find optimal solutions to uncertain information, for data from real sensors that can contain offsets and drifts. For a more detailed comparison of different methods to analyze IMU data, a number of recommendable surveys are available (Filippeschi et al. 2017; Bergamini et al. 2014).

The information from an IMU supports the analysis of a "well-defined" problem. For each moment in time t_i, the IMU provides six parameters: the three components of $\boldsymbol{\omega}(t_i)$, plus the three components of $\mathbf{acc}(t_i)$. The analysis must resolve six variables: the three components of the position vector $\mathbf{x}$, and the three parameters defining the orientation, e.g., the quaternion vector $\mathbf{q}$ (see Chap. 4). This poses a well-defined problem: there are equal numbers of input measurements and output variables. The solutions for orientation and position are presented in Sects. 6.2.1 and 6.2.2, respectively.

Note that since the IMU only provides information about the *derivatives* of position and orientation, the *initial values* for position, velocity, and orientation are required in order to find the unique position and orientation at time t_i.

6.2.1 Orientation in Space

Inertial sensors typically provide the linear acceleration **acc** and the angular velocity $\boldsymbol{\omega}$ of the sensor (in the sensor's coordinate system). However, these values do not directly provide the movement of the sensor with respect to the fixed-space coordinate system: these values are measured by the *moving* sensor, which means that one obtains linear acceleration and angular velocity in sensor coordinates.

In the upcoming equations, the following notation will be used: $\mathbf{x}^{\text{space}}$ is a vector expressed in space coordinates, and $\mathbf{x}^{\text{object}}$ the corresponding vector locally expressed with respect to the object.

If the following information is available

- the initial conditions $\mathbf{x}(t_0)$, $\mathbf{R}(t_0)$, $\mathbf{vel}(t_0)$,
- the IMU output $\mathbf{acc}^{\text{object}}(t_i)$ and $\boldsymbol{\omega}^{\text{object}}(t_i)$ in sensor coordinates, and
- the sampling period Δt, assuming that this sampling period is constant,

one can calculate the new position and orientation in space, assuming that the linear acceleration and angular velocity remain approximately constant during a time interval Δt.

Orientation measurements are independent of the linear acceleration, whereas the measured gravito-inertial acceleration (i.e., the output of the acceleration sensor) depends on the current orientation (see Eq. 2.7). Therefore, reconstruction of position and orientation from the sensor data has to start with the determination of object orientation. The orientation of an object is determined by its starting orientation and its angular velocity $\boldsymbol{\omega}^{\text{object}}$ as described in the following.

Let the object under investigation start with its orientation with respect to space given by the rotation matrix $\mathbf{R}^{\text{space}}_{\text{object,start}}$. The subsequent rotation of this object for a short duration Δt about a constant, space-fixed axis $\mathbf{n}$ with a constant velocity $\boldsymbol{\omega}^{\text{space}}$ is described by $\Delta\mathbf{R}^{\text{space}}$, which depends on the current axis of rotation ($\mathbf{n}$), the rotational speed ($\boldsymbol{\omega}^{\text{space}}$), and the time duration of sampling (Δt). Since in combined rotations the rotation applied first (here the starting orientation) is written on the right-hand-side of the matrix multiplication (see Sect. 3.4), the new orientation of the object in fixed-space coordinates is given by

$$\mathbf{R}^{\text{space}}_{\text{object,new}} = \Delta\mathbf{R}^{\text{space}} \cdot \mathbf{R}^{\text{space}}_{\text{object,start}}. \tag{6.16}$$

Note that the correct sequence of the matrices in this matrix multiplication is essential.

The next analysis step is critical for the correct reconstruction of the object orientation in space: if the angular acceleration is recorded in the sensor/object coordinates (e.g., from an inertial tracking device mounted on the object), it first has to be transformed from the object-fixed reference frame to a space-fixed reference frame. Let $\Delta\mathbf{R}^{\text{object}}$ describe the *movement* as seen in the object coordinates, and $\mathbf{R}^{\text{space}}_{\text{object}}$ the orientation of the object with respect to space. Then according to Eq. (A.6), which describes how a rotation matrix transforms for a change of the coordinate system, the movement with respect to space is given by

$$\Delta\mathbf{R}^{\text{space}} = \mathbf{R}^{\text{space}}_{\text{object}} \cdot \Delta\mathbf{R}^{\text{object}} \cdot \left(\mathbf{R}^{\text{space}}_{\text{object}}\right)^{-1}. \tag{6.17}$$

Inserting Eq. (6.17) into Eq. (6.16), and noting that for short durations $\mathbf{R}^{\text{space}}_{\text{object}} \approx \mathbf{R}^{\text{space}}_{\text{object,start}}$ leads to

$$\mathbf{R}^{\text{space}}_{\text{object,new}} = \mathbf{R}^{\text{space}}_{\text{object,start}} \cdot \Delta\mathbf{R}^{\text{object}} \cdot \left(\mathbf{R}^{\text{space}}_{\text{object,start}}\right)^{-1} \cdot \mathbf{R}^{\text{space}}_{\text{object,start}} = \mathbf{R}^{\text{space}}_{\text{object,start}} \cdot \Delta\mathbf{R}^{\text{object}}. \tag{6.18}$$

Comparing this to Eq. (6.16), we see that for incremental rotations the only thing that changes as we switch from an angular movement recorded with respect to space to an angular movement recorded with respect to the object is the sequence of the matrix multiplication!

For practical calculations, it is easiest to determine the orientation from the angular velocity using quaternions, as has been shown in Sect. 5.3.2. There angular velocities were used that describe the angular movement with respect to space $\boldsymbol{\omega}^{space}$, and the final orientation was given by Eq. (5.25). Now if instead angular velocities are used that have been measured with respect to the object $\boldsymbol{\omega}^{object}$, Eqs. (6.16) and (6.18) imply that the only thing that has to be changed is the sequence of rotations. Using quaternions, the final 3-D orientation of an object with respect to space whose orientation has been recorded with an IMU is, therefore, given by

$$\tilde{\boldsymbol{q}}(t) = \tilde{\boldsymbol{q}}(t_0) \circ \Delta\tilde{\boldsymbol{q}}^{object}(t_1) \circ \Delta\tilde{\boldsymbol{q}}^{object}(t_2) \circ \ldots \circ \Delta\tilde{\boldsymbol{q}}^{object}(t_{n-1}) \circ \Delta\tilde{\boldsymbol{q}}^{object}(t_n)\,, \quad (6.19)$$

where as in Eq. (5.25) the quaternion vectors are given by

$$\Delta\mathbf{q}^{object}(t_i) = \mathbf{n}(t)\sin\left(\frac{\Delta\phi(t_i)}{2}\right) = \frac{\boldsymbol{\omega}^{object}(t_i)}{\left|\boldsymbol{\omega}^{object}(t_i)\right|}\sin\left(\frac{\left|\boldsymbol{\omega}^{object}(t_i)\right|\Delta t}{2}\right) \quad (6.20)$$

with $\tilde{\boldsymbol{q}}(t_0)$ the starting quaternion, and $\Delta\tilde{\boldsymbol{q}}^{object}(t_i)$ the quaternions corresponding to the vector parts $\Delta\mathbf{q}^{object}(t_i)$.

6.2.2 *Position in Space*

Initial Orientation

As mentioned in Sect. 6.2.2, accelerometers measure the gravito-inertial acceleration, not just the acceleration caused by the movement of the object in space. In order to obtain only the linear acceleration of an object in space from the signals of an acceleration sensor, one first has to subtract gravity from the measured acceleration signal. For that, the orientation with respect to gravity first has to be determined from the gyroscope signals (see Fig. 6.5).

If one knows the initial position $\mathbf{x}(t=0)$ and the initial velocity $\left.\frac{d\mathbf{x}}{dt}\right|_{t=0}$, one can integrate the acceleration signals to obtain velocity and position in space.

If the sensor is rotated, the acceleration component contributed by gravity points in a direction that depends on the orientation of the sensor coordinate system with respect to space. Note that a rotation of the sensor *about* gravity does not change

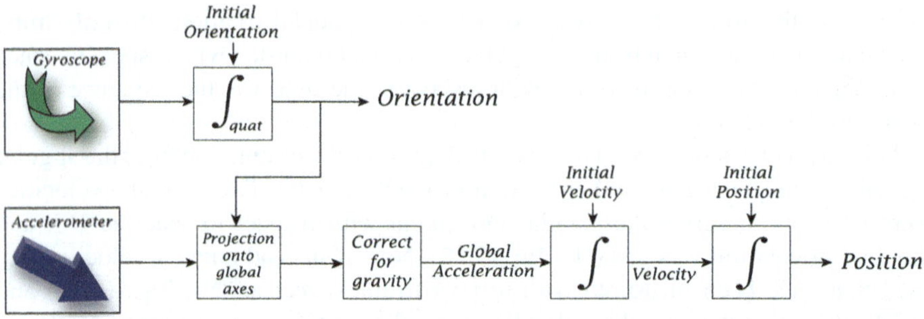

Fig. 6.5 Strapdown inertial navigation algorithm

the gravitational acceleration measured. In contrast, a tilt of the sensor with respect to gravity changes the output of the IMU accelerometer. The application example in Sect. 4.5.2 shows how to combine the knowledge about the approximate orientation of the sensor with the measured acceleration signal at $t = 0$, to obtain the best possible estimate of the initial sensor orientation.

Finding Acceleration, Velocity, and Position in Space

If the orientation of the sensor with respect to space is denoted by $\mathbf{R}_{\text{sensor}}^{\text{space}}$, then the measured direction of gravity is

$$\mathbf{g}^{\text{sensor}} = \left(\mathbf{R}_{\text{sensor}}^{\text{space}}\right)^{-1} \cdot \mathbf{g}^{\text{space}} \tag{6.21}$$

$$\mathbf{R}_{\text{sensor}}^{\text{space}} \cdot \mathbf{g}^{\text{sensor}} = \mathbf{g}^{\text{space}} \tag{6.22}$$

The movement of the sensor in space can be determined from Eq. (2.7). Seen from a space-fixed coordinate system

$$\mathbf{acc}_{\text{measured}}^{\text{space}} = \mathbf{R}_{\text{sensor}}^{\text{space}} \cdot \mathbf{acc}_{\text{measured}}^{\text{sensor}}, \tag{6.23}$$

the linear acceleration caused by movement in space is

$$\mathbf{acc}_{\text{movement}}^{\text{space}} = \mathbf{R}_{\text{sensor}}^{\text{space}} \cdot \mathbf{acc}_{\text{measured}}^{\text{sensor}} - \mathbf{g}^{\text{space}} \tag{6.24}$$

In many applications, one is interested in the position of the sensor (and thus of the object under investigation) with respect to space. When the inertial acceleration with respect to space is known, the positional change from the initial position can be found through integration.

Note that in the equations above, $\mathbf{acc}_{\text{movement}}^{\text{space}}$ is the acceleration component caused by the movement with respect to space (Eq. 6.24), not the acceleration indicated by the accelerometer!

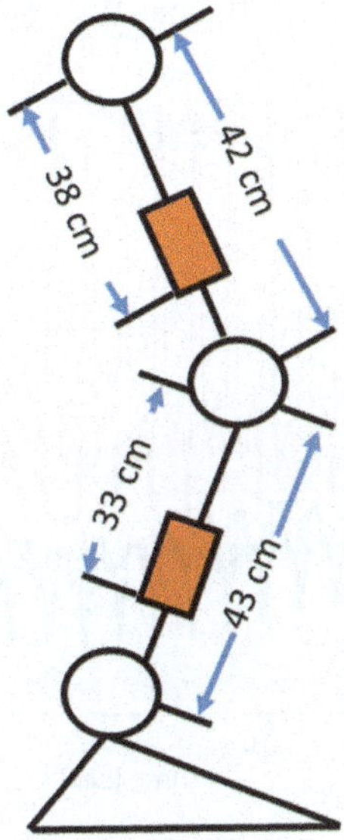

Fig. 6.6 Positioning of IMUs on the right leg during the movement recording

6.3 Applications: Gait Analysis

A simple analysis of leg movements while walking on a treadmill can demonstrate application of the utilities provided with `scikit-kinematics`. To record the movement of the upper and lower leg while walking on a treadmill, two IMUs from XSens were strapped to the upper and lower leg, as indicated in Fig. 6.6.

The following piece of code demonstrates how the orientation of the lower leg with respect to the upper leg can be calculated. Thereby, the approximate initial orientation of the IMUs is provided, since the magnetic field signals are not used for the analysis. This "knee-movement" is expressed with respect to the space-fixed coordinate system.

python™ **Code**: `C6_gait_analysis.py`: Demonstration of a quick evaluation of knee movements while walking on a treadmill (Fig. 6.7).

Listing 6.1: C6_gait_analysis.py

```
'''
Calculation of 3-D knee orientation from IMU-data
of upper- and lower-leg.

'''
# author: Thomas Haslwanter, date: Jan-2018, Ver: 1.1

# Import standard packages
import numpy as np
```

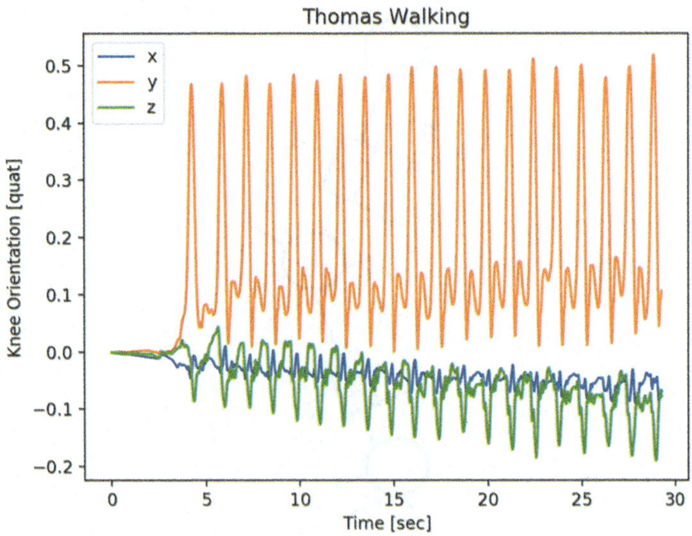

Fig. 6.7 3-D orientation of the knee while walking on a treadmill. (The data still contain small drift artifacts, which may be due to sensor slippage on the leg.)

```
import matplotlib.pyplot as plt
import os

# Import skinematics
from skinematics.sensors.xsens import XSens
from skinematics.quat import Quaternion

# Get the data
data_dir = r'D:\Users\thomas\Coding\Python\scikit-kinematics\
    skinematics\tests\data'
infile_ll = os.path.join(data_dir, 'walking_xsens_lowerLeg.
    txt')
infile_ul = os.path.join(data_dir, 'walking_xsens_upperLeg.
    txt')

# Provide the approximate initial orientation of the IMUs
initial_orientation = np.array([[0,0,-1], [1, 0, 0],
    [0,-1,0]]).T

sensor_ul = XSens(infile_ul, R_init=initial_orientation)
sensor_ll = XSens(infile_ll, R_init=initial_orientation)

# Convert the orientation to quaternions
q_upperLeg = Quaternion(sensor_ul.quat)
q_lowerLeg = Quaternion(sensor_ll.quat)

'''
```

```
Calculate the 3-D knee orientation, using "Quaternion"
    objects
Using the two rules for combined rotations
    * From right to left
    * From the inside out
we get that the orientation of the
    lower_leg = upper_leg * knee
Bringing the "upper_leg" to the other side, we have
    knee = inv(upper_leg) * lower_leg
'''
knee = q_upperLeg.inv() * q_lowerLeg


# Show the results
time = np.arange(len(knee)) / sensor_ul.rate
plt.plot(time, knee.values[:,1:])
plt.title('Thomas Walking')
plt.xlabel('Time [sec]')
plt.ylabel('Knee Orientation [quat]')
plt.legend(['x', 'y', 'z'])
plt.show()
```

6.4 Exercises

Exercise 6.1: An (Almost) Simple Rotation

Assume that the body-fixed coordinate system is such that the x-axis points forward, the y-axis to the left, and the z-axis up. Before the rotations, the space-fixed coordinate system and the body-fixed coordinate system coincide. Now two rotations are performed: the body-fixed IMU is read out at 100 Hz, and shows the angular velocities indicated in Fig. 6.8.

Try to answer the following questions:

- How can angular orientation be calculated—in principle—from angular velocity?
- If the cumulative rotation during the first second amounts to 45°, what are the unit vectors of the body-fixed coordinate system after 1 s?
- If the rotation during the 2nd second amounts to 30°, what are the unit vectors of the body-fixed coordinate system after 2 s?

Exercise 6.2: Pendulum

Consider the signals measured by an IMU that is attached to a perfect pendulum, with a length $l = 20$ cm.

Try to answer the following questions:

- What are position and orientation of the pendulum as a function of time, if the pendulum is released at $t = 5$ s, at a deflection of 5°?
- What are the corresponding values if the initial deflection is 70°?
- Which acceleration signals are recorded during the movement?

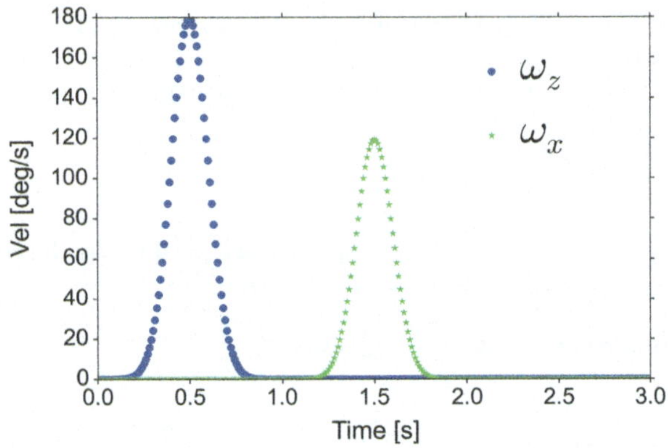

Fig. 6.8 Rotation about the z- and the x-axis

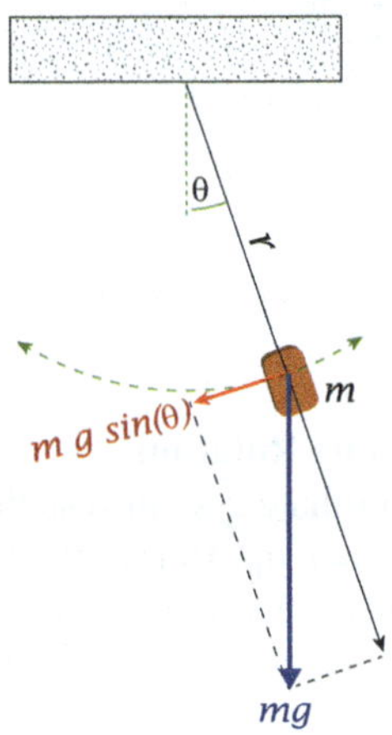

Fig. 6.9 IMU attached to a pendulum

- Which angular velocity signals are recorded during the movement?
- Do the expected measurement signals meet your expectation?

The solution to the following exercise shows that it can be almost impossible to guess the spatial movement from the accelerometer data alone.

python **Code**: `C6_examples_IMU.py`: Example of working with data from IMU sensors (p. 153) (Fig. 6.9).

Chapter 7
Multi-sensor Integration

Under ideal conditions, the algorithms described above work perfectly. From the initial orientation plus the gyroscope data, they can first determine the orientation. And knowing orientation, they can cancel out the contribution from gravity, and—given the initial position and velocity—find the current position in space.

However, the analysis of real sensory signals is not quite that simple. Real sensory signals include a number of artifacts like offsets and drifts (Woodman 2007). And since offsets lead to velocity errors that grow linearly with time, and position errors grow quadratically (as described in Sect. 2.2.2), the analytical solutions rarely get applied directly.

But a wealth of algorithms exist, using different approaches to deal with gyroscope bias drift, inertial acceleration, and magnetic disturbances (e.g. Mahony et al. 2008; Savage 2006; Sabatini 2006; Roetenberg et al. 2007). Two main sensor fusion approaches have been proposed: stochastic filtering, often implemented in the form of an extended Kalman filter. And the so-called "complementary filtering" approaches, which fuse multiple noisy measurements from the gyroscopes, accelerometers, and magnetometers that have complementary spectral characteristics. For each measurement, the complementary filtering uses only the part of the signal frequency spectrum that contains useful information. (This is reflected in the name, *complementary filters*.) Unfortunately, due to the varying conventions used in the different publications, such as quaternions, Euler angles, rotation matrices, and rotations of objects versus rotations of coordinate systems, direct comparisons of the different approaches are often difficult.

This chapter first provides an introduction to working with uncertain data. After that the principle of Kalman filters is introduced. In the last section, an example of a complementary filter that has received much attention for the evaluation of IMU data is presented, the filter proposed by Madgwick et al. (2011).

T. Haslwanter, *3D Kinematics*, https://doi.org/10.1007/978-3-319-75277-8_7

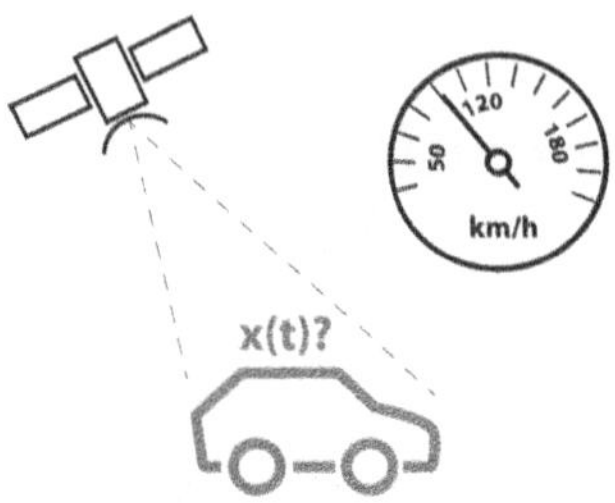

Fig. 7.1 How best to combine the information from a GPS and the speedometer, to obtain the optimal estimate of the current position?

7.1 Working with Uncertain Data

As[1] an example application, consider the problem of determining the precise location of a car (Fig. 7.1). The car can be equipped with a GPS unit that provides an estimate of the position within a few meters. The GPS estimate is likely to be noisy; readings "jump around" rapidly, though always remaining within a few meters of the real position. In addition, since the car is expected to follow the laws of physics, its position can also be estimated by integrating its velocity over time, determined by keeping track of wheel revolutions and the angle of the steering wheel. This is a technique known as "dead reckoning". Typically, the dead reckoning will provide a very smooth estimate of the car's position, but it will drift over time as small errors accumulate.

What makes this process particularly challenging is that neither the current position/velocity of the car nor the GPS measurement are 100% accurate. To prepare the mathematical ground for working with probabilities, the next section will start with an introduction on how *uncertain information* can be described mathematically.

7.1.1 Uncertain Data in One Dimension

Normal Distribution

For simplicity, we start out with a one-dimensional, uncertain piece of information: a one-dimensional position measurement. The measurement indicates a certain value μ, but also has an uncertainty σ (Fig. 7.2). The probability that the value of a measurement is correct is characterized by a so-called "probability distribution". In many cases, this probability distribution is well described by a "normal probability distribution", also called a "Gaussian probability distribution":

$$\mathcal{N}(x \mid \mu, \sigma) = \frac{1}{\sqrt{2\sigma^2\pi}} e^{-\frac{(x-\mu)^2}{2\sigma^2}} \tag{7.1}$$

[1]This section is strongly based on the presentation http://www.bzarg.com/p/how-a-kalman-filter-works-in-pictures/ by Timm Babb.

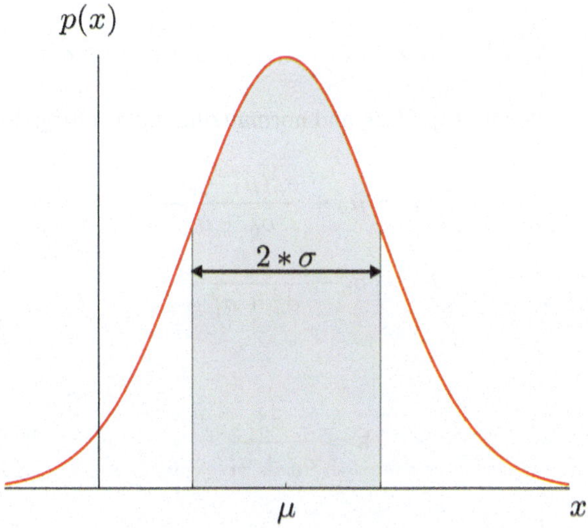

Fig. 7.2 Normal distribution, centered about μ, with a standard deviation of σ

where μ is the mean or expected value of the distribution, and σ is the standard deviation (σ^2 is the variance).

Combination of Two Normal Distributions

How can one obtain the best estimate for the position of the car if one has two different measurements, in our example the prediction from the dead reckoning and the GPS measurement? Luckily, the product of two Gaussians is again a Gaussian (Fig. 7.3). Let μ_i be the best guess of the measurement i, and σ_i the corresponding standard deviation. Then, the combined probability distribution can then be obtained with

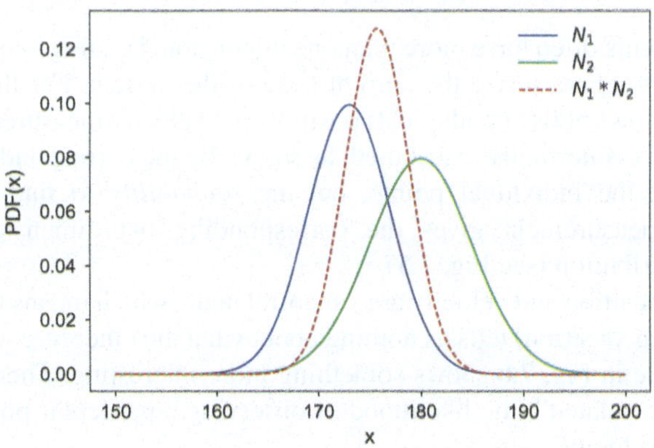

Fig. 7.3 The product of two normal probability distributions (solid lines) is again normally distributed (dotted line)

$$\mathcal{N}(x \mid \mu_0, \sigma_0) \cdot \mathcal{N}(x \mid \mu_1, \sigma_1) \sim \mathcal{N}(x \mid \mu', \sigma') \tag{7.2}$$

Substituting Eq. (7.1) into Eq. (7.2), and normalizing the resulting distribution gives

$$\mu' = \mu_0 + \frac{\sigma_0^2(\mu_1 - \mu_0)}{\sigma_0^2 + \sigma_1^2} \tag{7.3}$$

$$\sigma'^2 = \sigma_0^2 - \frac{\sigma_0^4}{\sigma_0^2 + \sigma_1^2}. \tag{7.4}$$

With

$$\mathbf{k} = \frac{\sigma_0^2}{\sigma_0^2 + \sigma_1^2} \tag{7.5}$$

we obtain

$$\begin{aligned} \mu' &= \mu_0 + \mathbf{k}(\mu_1 - \mu_0) \\ \sigma'^2 &= (1 - \mathbf{k})\sigma_0^2. \end{aligned} \tag{7.6}$$

The variable **k** in Eq. (7.6) corresponds to the "Kalman Gain" of the Kalman filter described in the next section, and the combination of two probability distributions is equivalent to the action *Update* in Fig. 7.8, since the information from one system is combined with that from another system.

7.1.2 Uncertain Data in Multiple Dimensions

In practice systems often have more than one dimension, i.e., they require more than one parameter to characterize the current state of the system. For the car example here, Fig. 7.4 shows 500 (hypothetical) position and velocity measurements. Each of these parameters is normally distributed, as shown by the corresponding histogram.

Plotting not the individual points, but the *probability* to find a given position/velocity measurement gives the corresponding two-dimensional Gaussian probability distribution (see Fig. 7.5).

In Fig. 7.5, position and velocity are "uncorrelated", which means that the state of one variable (e.g. position) tells us nothing about what the other (e.g. velocity) might be. The example in Fig. 7.6 shows something more interesting. There position and velocity are "correlated": the likelihood of observing a particular position depends on the current velocity.

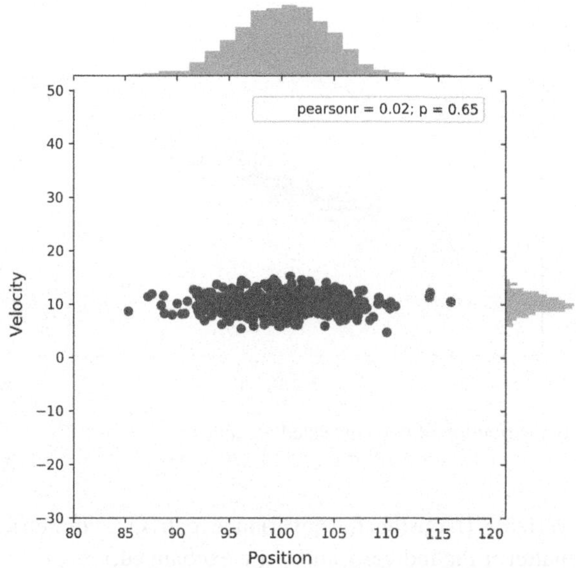

Fig. 7.4 500 samples from uncorrelated position and velocity measurements. The projections on the top and on the right show the corresponding sample histograms

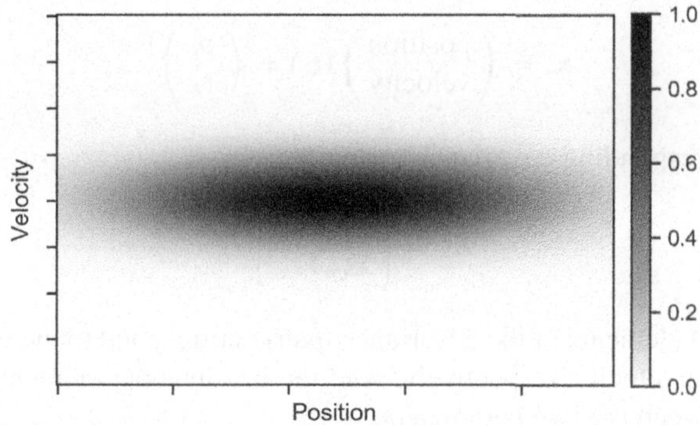

Fig. 7.5 Probability distribution of two uncorrelated variables. The colorbar on the right side shows the scale how likely it is to obtain a measurement at any given location

This kind of situation might arise if, for example, the estimate of a new position is based on an old one. If the velocity was high, the car probably moved farther, so the new position will be more distant. If the car drove slowly, it did not get as far.

This kind of relationship is really important to keep track of, because it provides more information: one measurement contains information about what the other could be. This correlation is captured by the so-called "covariance matrix" $\mathbf{\Sigma}$. Each element Σ_{ij} of the matrix quantifies the degree of correlation between the ith state variable

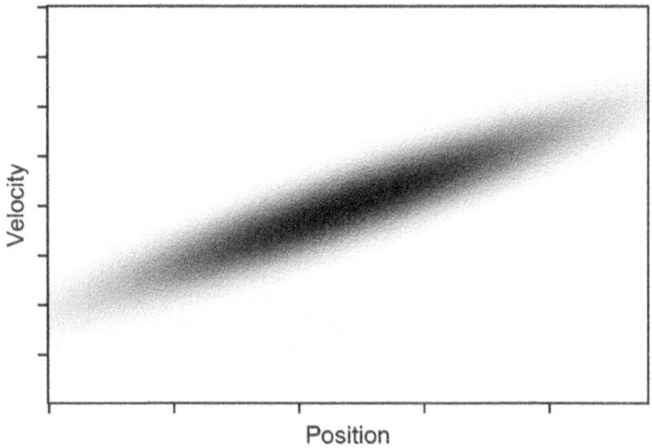

Fig. 7.6 Probability distribution of two correlated variables

and the jth state variable. (Note that the covariance matrix is symmetric, which means that it does not matter if the indices i and j are exchanged.)

To make the notation more concise, let p_k be the position at time t_k, and v_k the corresponding velocity. The "state vector" describing the object is now given by the vector $\mathbf{x}$, defined as

$$\mathbf{x}_k = \begin{pmatrix} \text{position} \\ \text{velocity} \end{pmatrix} (t_k) = \begin{pmatrix} p_k \\ v_k \end{pmatrix}. \tag{7.7}$$

And the corresponding covariance matrix is

$$\mathbf{\Sigma}_k = \begin{bmatrix} \Sigma_{pp} \, \Sigma_{pv} \\ \Sigma_{vp} \, \Sigma_{vv} \end{bmatrix}. \tag{7.8}$$

The diagonal elements of the covariance matrix correspond to the *variance* of the position and the velocity, respectively. And the off-diagonal elements quantify the *correlation* between the two parameters.

The combination of uncertain multidimensional measurements with uncertain state expectations requires matrix versions of Eqs. (7.5) and (7.6).

If $\mathbf{\Sigma}$ is the covariance matrix of a Gaussian blob, and the vector $\boldsymbol{\mu}$ its mean along each axis, then the Kalman gain matrix $\mathbf{K}$ is

$$\mathbf{K} = \mathbf{\Sigma}_0 \cdot (\mathbf{\Sigma}_0 + \mathbf{\Sigma}_1)^{-1}, \tag{7.9}$$

and the new mean $\boldsymbol{\mu}'$ and the new covariance matrix Σ' are (Fig. 7.7)

$$\begin{aligned} \boldsymbol{\mu}' &= \boldsymbol{\mu}_0 + \mathbf{K} \cdot (\boldsymbol{\mu}_1 - \boldsymbol{\mu}_0) \\ \mathbf{\Sigma}' &= (1 - \mathbf{K}) \cdot \mathbf{\Sigma}_0. \end{aligned} \tag{7.10}$$

The subscripts in Eqs. (7.9) and (7.10) refer to the first and second set of measurements.

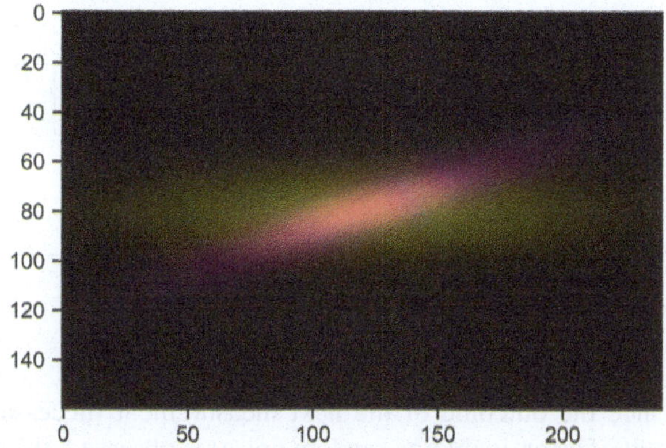

Fig. 7.7 The combination from the information from two higher dimensional probability distributions reduces the uncertainties in our estimates: the bright area indicates where both measurement_0 (magenta) and measurement_1 (yellow) agree

7.2 Kalman Filter

7.2.1 Idea Behind Kalman Filters

Kalman[2] filtering is an algorithm that uses a series of measurements observed over time, containing statistical noise and other inaccuracies, and produces estimates of unknown variables that tend to be more precise than those based on a single measurement alone, by using Bayesian inference and estimating a joint probability distribution over the variables for each time frame. The elegant feature of the Kalman filter is that the uncertainty in the data is taken into consideration, and the maximum amount of knowledge is extracted from the given information. The filter is named after Rudolf E. Kálmán (1930–2016), one of the primary developers of its theory.

The Kalman filter has numerous applications in technology. A common application is for guidance, navigation, and control of vehicles, particularly aircraft and spacecraft. Furthermore, the Kalman filter is a widely applied concept in time series analysis used in fields such as signal processing and econometrics. Kalman filters are also one of the main topics in the field of robotic motion planning and control, and they are sometimes included in trajectory optimization. In neuroscience, the Kalman filter has found use in modeling the central nervous system's control of movement. Due to the time delay between issuing motor commands and receiving sensory feedback, use of the Kalman filter provides the needed model for making estimates of the current state of the motor system and issuing updated commands (Wolpert and Ghahramani 2000).

The algorithm works in a two-step process (see Fig. 7.8). In the *Prediction* step, the Kalman filter produces estimates of the current state variables, along with their

[2]This section is taken from https://en.wikipedia.org/wiki/Kalman_filter

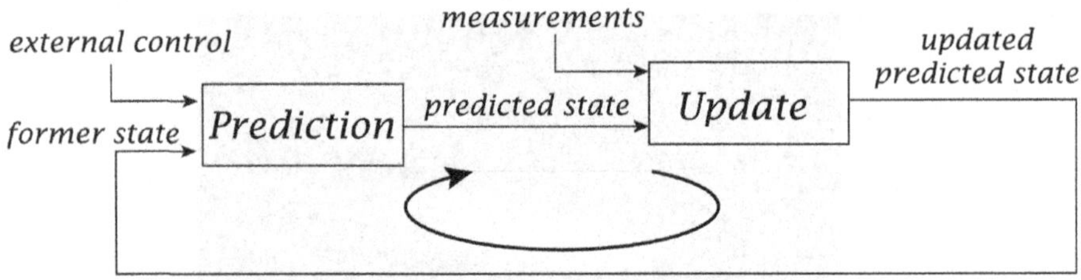

Fig. 7.8 Simplified diagram describing the iterative update of a system

uncertainties. Once the outcome of the next measurement (necessarily corrupted with some amount of error, including random noise) is observed, the state estimates are combined with the measurements in an *Update* step using a weighted average, with more weight being given to estimates with higher certainty. The algorithm is recursive. It can run in real time, using only the present input measurements and the previously calculated state and its uncertainty matrix; no additional past information is required.

The Kalman filter does not require any assumption that the errors are Gaussian. However, the filter yields the exact conditional probability estimate in the special case when all errors are Gaussian distributed.

For Kalman filters, the underlying model is a Bayesian model similar to a "hidden Markov model", but where the state space of the latent variables is continuous and where all latent and observed variables have Gaussian distributions.

Example Application

In the example with the car on a highway, the parameters (position/velocity) describe the current state of this system, and are sampled at equal time increments Δt. Knowing the *former state* of the system (position/velocity), and the *external control* (position of the accelerator pedal), one can calculate the *predicted state* of the system Δt seconds later (Fig. 7.8). Since the car may have encountered, e.g., a steep, bad section of the road, the *predicted state* will not match up exactly with the *measurements* (e.g., the position/velocity information from the GPS signals). In order to get the best possible estimate of the current position, these two pieces of information are integrated in the *Update*, to get the *updated predicted state*. This is the new starting point, and the process begins all over again.

In this example, the Kalman filter can be thought of as operating in two distinct phases: *Prediction* and *Update*. In the prediction phase, the car's old position will be modified according to the physical laws of motion (the dynamic or "state transition" model) plus any changes produced by the accelerator pedal and steering wheel. Not only will a new position estimate be calculated, but a new covariance will be calculated as well, providing information about the uncertainty of the car's position. Perhaps the covariance is proportional to the speed of the car because we are more uncertain about the accuracy of the dead reckoning position estimate at high speeds but very certain about the position estimate when moving slowly. Next, in the update phase, a measurement of the car's position is taken from the GPS unit. Along with this measurement comes some amount of uncertainty, and its covariance relative to that of the prediction from the previous phase determines how much the new measurement will affect the updated prediction. Ideally, if the dead reckoning estimates tend to drift away from the real position, the GPS measurement should pull the position estimate back toward the real position but not disturb it to the point of becoming rapidly changing and noisy.

7.2.2 *State Predictions*

How can state predictions, corresponding to the box *Prediction* in Fig. 7.8, be implemented? In the following first the equations without *external control* will be considered, and then external input will be added. In order to facilitate the overview, and the correspondence between the equations and Figs. 7.9 and 7.10, the elements in the following equations will use the same colors as the corresponding elements in the figures.

Without External Control

In the example the current position/velocity is known at time t_{k-1} (Fig. 7.9, left). If one wants to know the best estimate for the position/velocity at time t_k (Fig. 7.9, right), one can write that down as

$$\begin{aligned} p_k &= p_{k-1} + \Delta t * v_{k-1} \\ v_k &= v_{k-1}. \end{aligned} \tag{7.11}$$

Using vector and matrix notation, this can be written as

$$\mathbf{x}_k = \mathbf{F}_k \cdot \mathbf{x}_{k-1}. \tag{7.12}$$

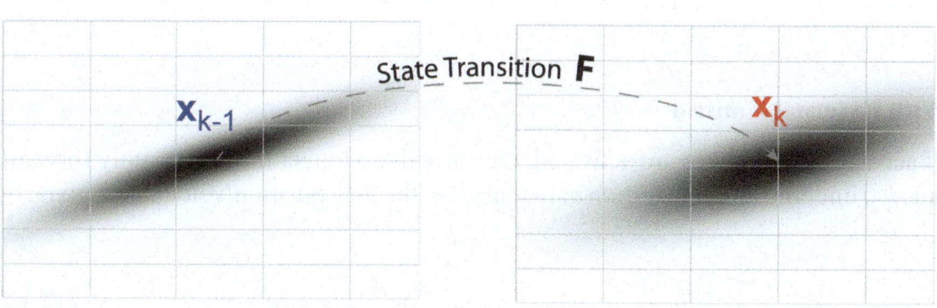

Fig. 7.9 Knowledge of the previous state (left, $\mathbf{x}_{k-1}$) and the state-transition matrix **F** allows calculation of the new state (right, $\mathbf{x}_k$)

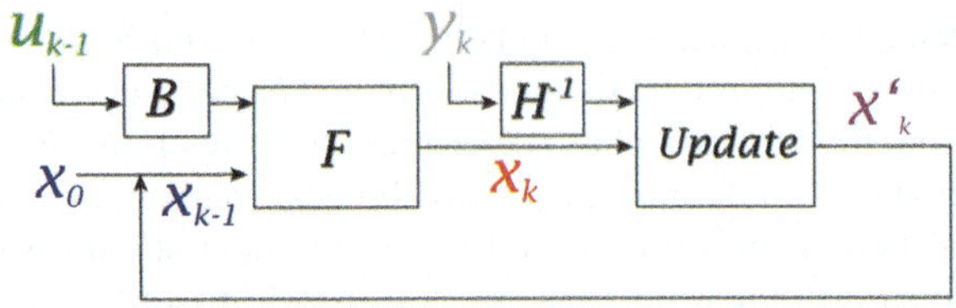

Fig. 7.10 Simplified scheme describing the iterative update of a system, taking external factors $\mathbf{u}_{k-1}$ and measurements y_k into consideration

Borrowing the terminology from the theory of control systems

- $\mathbf{x}_k$ is the *state* at time k, containing all the parameters required to describe the current state of system. (In our example, these are position and velocity, and $\mathbf{x}_0$ provides the initial state vector.)

$$\mathbf{x}_k = \begin{pmatrix} p_k \\ v_k \end{pmatrix}.$$

- $\mathbf{F}$ is the *state transition model*, in our case given by the matrix

$$\mathbf{F} = \begin{bmatrix} 1 & \Delta t \\ 0 & 1 \end{bmatrix}$$

which is applied to the previous state $\mathbf{x}_{k-1}$ in order to get the estimated new state $\mathbf{x}_k$.
- In the context of Kalman filters, the covariance matrix for the state $\mathbf{x}_{k-1}$ is typically indicated with $\mathbf{P}_{k-1}$

$$\mathbf{P} = \begin{bmatrix} \sigma_p^2 & \sigma_{pv} \\ \sigma_{vp} & \sigma_v^2 \end{bmatrix}.$$

If one knows how the *state transition* affects each individual point (Fig. 7.9), one can also calculate how the probability distribution develops. Elementary linear algebra gives

$$\mathbf{P}_k = \mathbf{F}_k \cdot \mathbf{P}_{k-1} \cdot \mathbf{F}_k^T. \tag{7.13}$$

With External Control

If accelerator or brakes are activated, the car will no longer move smoothly forward, but also undergo an acceleration a. In that case, the new position/velocity are given by

$$\begin{aligned} p_k &= p_{k-1} + \Delta t * v_{k-1} + \frac{1}{2} * a * \Delta t^2 \\ v_k &= v_{k-1} + a * \Delta t. \end{aligned} \tag{7.14}$$

Writing these equations in matrix form one obtains

$$\begin{aligned} \mathbf{x}_k &= \mathbf{F}_k \cdot \mathbf{x}_{k-1} + \begin{pmatrix} \frac{\Delta t^2}{2} \\ \Delta t \end{pmatrix} * a \\ &= \mathbf{F}_k \cdot \mathbf{x}_{k-1} + \mathbf{B}_k \cdot \mathbf{u}_k \end{aligned} \tag{7.15}$$

where:

- **B** is called the *control matrix*
- the *control vector* $\mathbf{u}_k$ (here a simple scalar a) characterizes the external input.

Also the external control is not 100% precise, but also has some variability. Denoting this external variability with **Q**, Eq. (7.13) turns into

$$\mathbf{P}_k = \mathbf{F}_k \cdot \mathbf{P}_{k-1} \cdot \mathbf{F}_k^T + \mathbf{Q}_k. \tag{7.16}$$

In words, with the knowledge of the covariance matrix for the previous state vector, $\mathbf{P}_{k-1}$, the state transition matrix $\mathbf{F}_k$, and the variability in external control, $\mathbf{Q}_k$, the covariance matrix for the new state can be calculated.

To conclude the update cycle, one final step has to be considered: external *measurements* (see Fig. 7.10, $\mathbf{y}_k$):

- $\mathbf{y}_k$, the observation at time k, is combined with the estimated state $\mathbf{x}_k$ to form a new estimate, $\mathbf{x}'_k$.
- With this new best estimate, the whole process is then repeated.

The next section shows how external measurements are included in the equations.

7.2.3 Measurements and Kalman Equations

Several sensors might provide information about the state of the system. For the time being it does not matter what they measure; perhaps one reads position and the other reads velocity. Each sensor says something indirect about the state. In other words, the sensors operate on a state and produce a set of readings. In the context of Kalman filters, it is typically assumed that the expected sensor signal is related to the state estimate $\mathbf{x}_k$ through a linear transformation

$$\mathbf{sensor} = \mathbf{H}_k \cdot \mathbf{x}_k. \tag{7.17}$$

And the uncertainty in the state estimate **P** propagates into an uncertainty in the sensor space via

$$\Sigma_{\text{expected}}^{\text{sensor}} = \mathbf{H}_k \cdot \mathbf{P}_k \cdot \mathbf{H}_k^T. \tag{7.18}$$

In Kalman equations, the mean sensor signal is typically labelled $\mathbf{z}_{k-1}$, and the *sensor noise* (i.e. the covariance of the sensor readings) with $\mathbf{R}_{k-1}$.

Putting It All Together

We have two distributions: the predicted measurement with

$$(\mu_0, \boldsymbol{\Sigma}_0) = (\mathbf{H}_k \cdot \mathbf{x}_k, \mathbf{H}_k \cdot \mathbf{P}_k \cdot \mathbf{H}_k^T)$$

and the observed measurement with

$$(\mu_1, \boldsymbol{\Sigma}_1) = (\mathbf{z}_k, \mathbf{R}_k).$$

Plugging these into Eq. (7.10) to find their overlap, we get

$$\begin{aligned} \mathbf{H}_k \cdot \mathbf{x}'_k &= \mathbf{H}_k \cdot \mathbf{x}_k + \mathbf{K} \cdot (\mathbf{z}_k - \mathbf{H}_k \cdot \mathbf{x}_k) \\ \mathbf{H}_k \cdot \mathbf{P}'_k \cdot \mathbf{H}_k^T &= (1 - \mathbf{K}) \cdot \mathbf{H}_k \cdot \mathbf{P}_k \cdot \mathbf{H}_k^T \end{aligned} . \tag{7.19}$$

And from Eq. (7.9), the Kalman gain is

$$\mathbf{K} = \mathbf{H}_k \cdot \mathbf{P}_k \cdot \mathbf{H}_k^T \cdot \left(\mathbf{H}_k \cdot \mathbf{P}_k \cdot \mathbf{H}_k^T + \mathbf{R}_k\right)^{-1}. \tag{7.20}$$

We can eliminate $\mathbf{H}_k$ off the front of every term in Eqs. (7.19) and (7.20) (note that one is hiding inside $\mathbf{K}$), and an $\mathbf{H}_k^T$ off the end of all terms in the equation for $\mathbf{P}'$ (Fig. 7.11).

$$\begin{aligned} \mathbf{x}'_k &= \mathbf{x}_k + \mathbf{K}' \cdot (\mathbf{z}_k - \mathbf{H}_k \cdot \mathbf{x}_k) \\ \mathbf{P}'_k &= \mathbf{P}_k - \mathbf{K}' \cdot \mathbf{H}_k \cdot \mathbf{P}_k \end{aligned} \tag{7.21}$$

$$\mathbf{K}' = \mathbf{P}_k \cdot \mathbf{H}_k^T \cdot (\mathbf{H}_k \cdot \mathbf{P}_k \cdot \mathbf{H}_k^T + \mathbf{R}_k)^{-1} \tag{7.22}$$

... giving us the complete equations for the *Update* step.

And that's it! $\mathbf{x}'$ is the new best estimate, and can be fed (along with $\mathbf{P}'_k$) back into another round of *Prediction* and *Update*.

These equations represent any linear system accurately. And for an implementation, of all the math above only Eqs. (7.13) and (7.16), (7.21), and (7.22) are required.

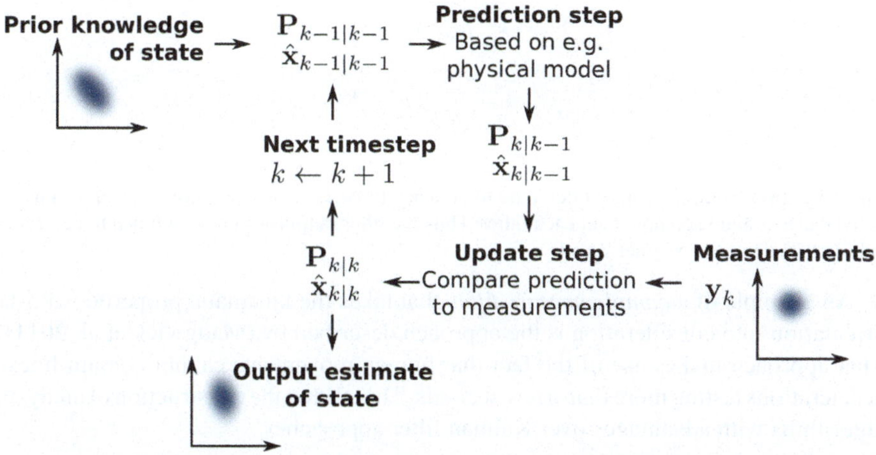

Fig. 7.11 The Kalman filter keeps track of the estimated state of the system and the variance or uncertainty of the estimate. The estimate is updated using a state transition model and measurements. $\mathbf{x}_{k|k-1}$ denotes the estimate of the system's state at time step k before the kth measurement $\mathbf{y}_k$ has been taken into account; $\mathbf{P}_{k|k-1}$ is the corresponding uncertainty (from Wikipedia)

7.2.4 *Kalman Filters with Quaternions*

For nonlinear systems the math gets more complicated, and methods like *extended Kalman filters* and *unscented Kalman filters* have to be used. These work in principle by linearizing the predictions and measurements about their mean. These extensions are particularly important for 3-D kinematics, since there the underlying algorithms are clearly nonlinear: for example, expressed with quaternions, the combination of two rotations requires a cross product calculation (Eq. 4.6).

The quaternion-based extended Kalman filter developed by Yun and Bachman for human body motion tracking (Yun and Bachmann 2006) is implemented in the `scikit-kinematics` package `imus`.

7.3 Complementary Filters

The "complementary filter" is a somewhat different, simple estimation technique that was developed in the flight control industry to combine measurements (Higgins 1975). This filter is actually a steady-state Kalman filter for a certain class of filtering problems. It does not consider any statistical description of the signal, but instead considers how x and y, two noisy measurements of some signal z, can be used to produce an estimate of the signal, $\hat{z}$, if the filter characteristics of the two measurements complement each other (see Fig. 7.12).

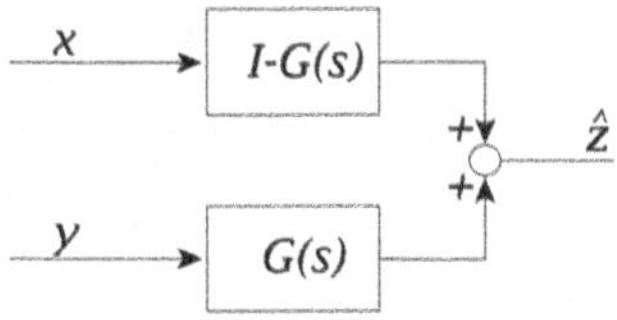

Fig. 7.12 In a "complementary filter", the filter characteristics of two measurements of a signal z, here labeled x and y, complement each other. Thus the filter outputs can be combined to achieve a better estimate of the original signal

An example of a *complementary filter* that takes the kinematic properties of 3-D orientation into consideration is the approach described by (Madgwick et al. 2011). That approach makes use of the fact that human movements cannot contain linear accelerations lasting more than a few seconds. This allows the construction of analysis algorithms with advantages over Kalman filter approaches.

7.3.1 *Gradient Descent Approach*

The algorithm for sensor integration developed by Madgwick is computationally very efficient (Madgwick et al. 2011). It uses a quaternion representation, allowing accelerometer and magnetometer data to be used in a "gradient descent algorithm" to compute the direction of the gyroscope measurement error as a quaternion derivative. The algorithm achieves levels of accuracy matching that of the Kalman-based algorithms. Open-source implementations of this algorithm are available for C, C#, and Matlab,[3] and for Python.[4]

The idea behind the *gradient descent* method is illustrated in Fig. 7.13: on an "error-surface", walk in the steepest downward direction (="gradient") in order to get to bottom most quickly.

The smart algorithm by Madgwick uses the following assumptions:

- On average, gravity points downward. This provides the direction of the space-fixed z-axis. And the horizontal component of the local magnetic field can be taken as the direction of the x-axis.
- Knowing the 3-D angular velocity ω_{t-1}, one can use a modification of Eq. (5.13),

$$\frac{d\tilde{q}}{dt} = \frac{1}{2}\tilde{\omega} \circ \tilde{q} \tag{7.23}$$

to get an estimate of the current orientation

$$\tilde{q}_{\omega,t} \approx \tilde{q}_{\omega,t-1} + \left(\frac{d\tilde{q}}{dt}\right)_{\omega,t} * \Delta t. \tag{7.24}$$

---

[3]http://x-io.co.uk/open-source-imu-and-ahrs-algorithms

[4]http://work.thaslwanter.at/skinematics/html/imus.html

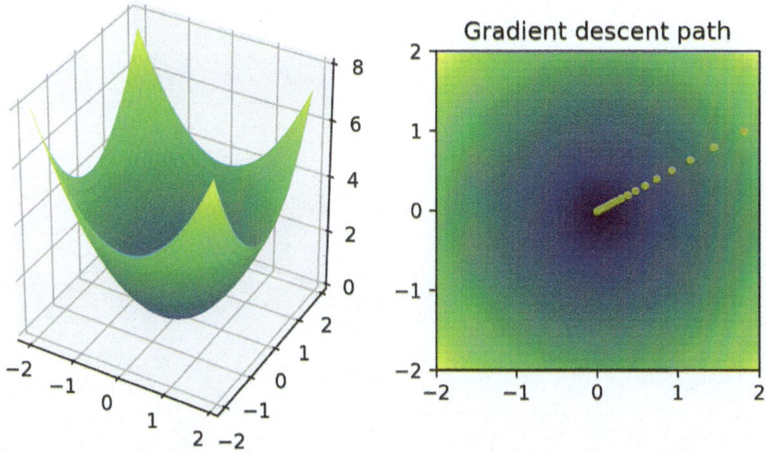

Fig. 7.13 (Left) Magnitude of deviation from gravity, forming a quadratic bowl. (Right) The yellow dots indicate the stepwise "gradient descent" to the bottom of a quadratic bowl, here starting at (x, y) = *(1.8/1.0)*

From the point of view of complementary filters, this provides the high-frequency input.

In Eq. (7.23), one has to be very careful with the sequence: Madgwick defines $\tilde{\boldsymbol{q}}$ to represent the orientation of the earth relative to the sensor, which results in the inversion of the sequence in Eq. (7.23)!

- Angular velocity sensors in IMUs typically show a substantial amount of drift and noise. In order to compensate for the resulting cumulative error, Madgwick combines this orientation estimate with a second estimate of orientation: interpreting the readout of the accelerometers as gravity, and knowing the direction of the local magnetic field, provides in combination a second orientation estimate, in addition to the one in Eq. (7.24). Here Madgwick proposes—based on investigations of the underlying kinematics—to perform *the first step of a gradient descent* into that direction, with a fixed magnitude which is set to compensate the typical gyroscope errors. Note that the step width for this step was adjusted in (Madgwick et al. 2011) to optimize the filter properties for the frequently used *XSens* sensor. From the point of view of complementary filters, this compensates for low-frequency drifts and errors.
- Since the local surroundings can have a significant effect on the local magnetic field, Madgwick's algorithm is modified such that this can only lead to errors in the heading direction, and no tilt errors.

The decision which filter is "best" for a given application depends on the specific application requirements. The filter by Madgwick was designed specifically for real-time implementations of human movement recordings. For other applications, for example, aerospace or in the automotive area, assumptions inherent in the approach by Madgwick may not hold.

Appendix A
Appendix—Mathematics

A.1 Mathematical Basics

A.1.1 Scalar Product

The "scalar product" of two vectors **a** and **b** is defined as

$$\begin{pmatrix} a_x \\ a_y \\ a_z \end{pmatrix} \cdot \begin{pmatrix} b_x \\ b_y \\ b_z \end{pmatrix} = a_x b_x + a_y b_y + a_z b_z = |\mathbf{a}| \cdot |\mathbf{b}| \cdot \cos(\theta)\,. \tag{A.1}$$

The geometric interpretation of the scalar product is the projection of one vector onto another (Fig. A.1, left).

A.1.2 Cross Product

The "cross product" or "vector product" of two vectors **a** and **b** is defined as

$$\begin{pmatrix} a_x \\ a_y \\ a_z \end{pmatrix} \times \begin{pmatrix} b_x \\ b_y \\ b_z \end{pmatrix} = \begin{pmatrix} a_y b_z - a_z b_y \\ a_z b_x - a_x b_z \\ a_x b_y - a_y b_x \end{pmatrix}\,. \tag{A.2}$$

The resulting vector is perpendicular to **a** and **b**, and vanishes if **a** and **b** are parallel. The length of the vector is given by the area of the parallelogram spanned by the two vectors (Fig. A.1, right).

T. Haslwanter, *3D Kinematics*, https://doi.org/10.1007/978-3-319-75277-8

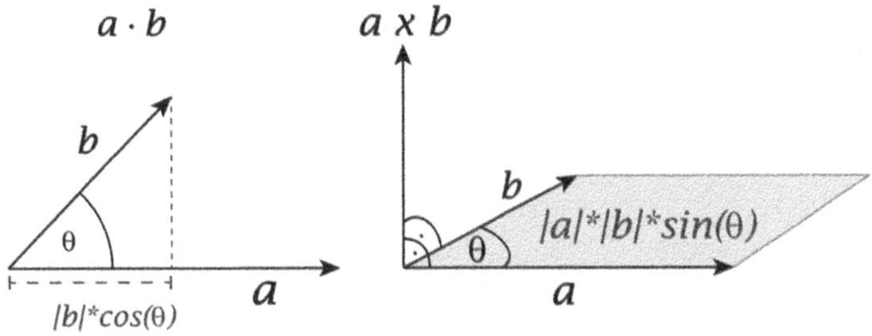

Fig. A.1 Graphical interpretation of scalar product (left) and vector product (right). Note that $|\mathbf{a}| = a$, and the same for b

It may be worth noting that the "invention" of the scalar and vector product goes back to quaternions: for two *pure* quaternions, the quaternion product separates neatly into the dot product for the scalar part, and the cross product for the vector part: with $\tilde{\boldsymbol{u}} = (0, \mathbf{u})$ and $\tilde{\boldsymbol{v}} = (0, \mathbf{v})$, Eq. (4.6) turns to

$$\tilde{\boldsymbol{u}} \circ \tilde{\boldsymbol{v}} = (\mathbf{u} \cdot \mathbf{v}) * 1 + (\mathbf{u} \times \mathbf{v}) \cdot \mathbf{I}\,. \tag{A.3}$$

A.1.3 Matrix Multiplication

In general, the multiplication of two matrices **A** and **B** is defined as

$$\mathbf{A} \cdot \mathbf{B} = \mathbf{C} \tag{A.4}$$

with $C_{ik} = \sum_j A_{ij} B_{jk}$.

This equation can also be used for multiplication of a matrix with a column vector, when the vector is viewed as a matrix with n rows and one column.

Example 1 (Rotation of a Point in Space) If $\mathbf{p}$ is a column vector indicating the position of a point in space, the location of the point rotated by the rotation matrix $\mathbf{R}$ is given by

$$\mathbf{p}' = \mathbf{R} \cdot \mathbf{p}\,. \tag{A.5}$$

Example 2 (Rotation of a Rotation Matrix) If $\mathbf{R}_{\text{rot}}$ describes a rotation about a space-fixed axis, a transformation of this rotation by the matrix $\mathbf{R}_{\text{trans}}$ is given by

$$\mathbf{R}'_{\text{rot}} = \mathbf{R}_{\text{trans}} \cdot \mathbf{R}_{\text{rot}} \cdot \mathbf{R}^{-1}_{\text{trans}}\,. \tag{A.6}$$

For example, if $\mathbf{R}_{\text{rot}}$ describes a positive rotation about the x-axis, and $\mathbf{R}_{\text{trans}}$ a rotation about the z-axis by 90°, then the result of Eq. (A.6), $\mathbf{R}'_{\text{rot}}$, describes a positive rotation about the y-axis. A simple computational proof can be implemented with Python:

```
# import the required packages
import numpy as np
import skinematics as skin

# Define a rotation about the x-axis by "alpha" ...
Rs_rot = skin.rotmat.R_s('x','alpha')
# ... and one about the z-axis by "theta"
Rs_trans = skin.rotmat.R_s('z', 'theta')

# Now "rotate the rotation", find the result
    for "theta"=90deg=pi/2, and show the result
Rs_rot_transformed = ( Rs_trans * Rs_rot
    * Rs_trans.inv() ).subs('theta', np.pi/2)
Rs_rot_transformed

# Output:
#Matrix(
#  [1.0*cos(alpha),      -6.1e-17*cos(alpha) + 6.1e-17,
    1.0*sin(alpha)],
#  [-6.1e-17*cos(alpha), 3.7e-33*cos(alpha) + 1.0,
    -6.1e-17*sin(alpha)],
#  [-1.0*sin(alpha),      6.1e-17*sin(alpha),
    cos(alpha)]] )
```

Ignoring numerical artifacts, which show up as values multiplied with factors $\epsilon < exp(-15)$, comparison of the output with Eq. (3.15) shows that the result is a rotation matrix for a positive rotation about the y-axis by α.

A.1.4 Basic Trigonometry

The basic elements of trigonometry are illustrated in Fig. A.2.

For small angles (i.e., $\theta << 1$) a Taylor series expansion of the sine, cosine, and tangent functions gives

$$\sin(\theta) \approx \tan(\theta) \approx \theta\,, \quad \text{and} \tag{A.7}$$

$$\cos(\theta) \approx 1 - \frac{\theta^2}{2}\,. \tag{A.8}$$

Thus for small angles, the change in the cosine is only a second-order effect.

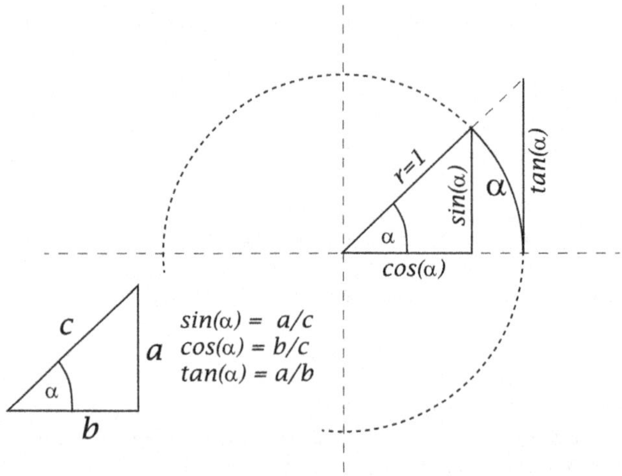

Fig. A.2 Basic trigonometry

A.2 Alternative Gram–Schmidt Calculation

The *Gram–Schmidt* algorithm can be used to calculate a right-handed orthogonal coordinate system, based on the location of three points $\mathbf{p}_i$ in space. One possible implementation has been given in Sect. 6.1.2. Another implementation is shown below:

$$\begin{aligned}\mathbf{a_x} &= \frac{\mathbf{p_1}-\mathbf{p_0}}{|\mathbf{p_1}-\mathbf{p_0}|}\,,\\ \mathbf{a_y} &= \frac{(\mathbf{p_{02}})_{\perp ax}}{|(\mathbf{p_{02}})_{\perp ax}|}\,,\ \text{with}\ (\mathbf{p_{02}})_{\perp \text{ax}} = (\mathbf{p_2}-\mathbf{p_0}) - (\mathbf{a_x}\cdot(\mathbf{p_2}-\mathbf{p_0})) * \mathbf{a_x}\,,\ \text{and}\\ \mathbf{a_z} &= \mathbf{a_x}\times\mathbf{a_y}\,.\end{aligned} \tag{A.9}$$

The second formula in Eq. (A.9) is obtained by looking at the decomposition of a vector $\mathbf{b}$ into one component parallel to $\mathbf{a}$, and one component perpendicular to $\mathbf{a}$: $\mathbf{b} = \mathbf{b}_{\|} + \mathbf{b}_{\perp}$ (see Fig. A.3), and utilizing the fact that $|\mathbf{a}| = 1$

$$\begin{aligned}\mathbf{b}_{\|} &= \mathbf{a} * |\mathbf{b}| * \cos(\alpha) = \mathbf{a} * (\mathbf{b}\cdot\mathbf{a})\\ \mathbf{b}_{\perp} &= \mathbf{b} - \mathbf{b}_{\|}\,.\end{aligned} \tag{A.10}$$

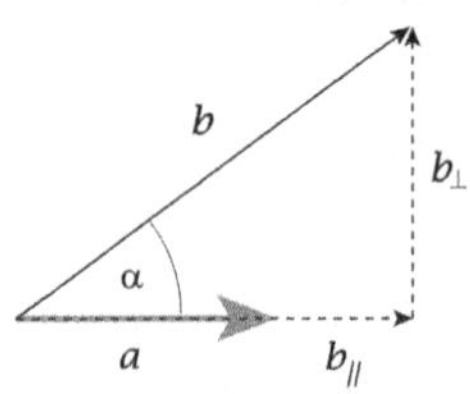

Fig. A.3 Decomposition of a vector **b** into two components: one parallel to **a** ($\mathbf{b}_{\|}$), and one perpendicular to **a** ($\mathbf{b}_{\perp}$)

The three resulting orthogonal unit vectors $\mathbf{a_i}(t)$ define the rotation matrix $\mathbf{R}$ which describes the orientation of an object in 3-D space:

$$\mathbf{R}(t) = \left[\, \mathbf{a_x}(t)\ \mathbf{a_y}(t)\ \mathbf{a_z}(t) \,\right] . \tag{A.11}$$

A.3 Proofs of Selected Equations

A.3.1 Quaternion Multiplication

Proof of Eq. (4.6) requires the calculation of $\sum_{i=0}^{3} q_i\, I_i * \sum_{j=0}^{3} p_j\, I_j$, using the commutation relationships specified in Eq. (4.5):

$$\begin{aligned}
\tilde{p} \circ \tilde{q} &= \sum_{i=0}^{3} p_i\, I_i * \sum_{j=0}^{3} q_j\, I_j \\
&= \left(q_0 * 1 + q_1 * \tilde{i} + q_2 * \tilde{j} + q_3 * \tilde{k}\right) * \left(p_0 * 1 + p_1 * \tilde{i} + p_2 * \tilde{j} + p_3 * \tilde{k}\right) \\
&= \left(q_0 p_0 * 1 * 1 + q_1 p_1 * \tilde{i} * \tilde{i} + q_2 p_2 * \tilde{j} * \tilde{j} + q_3 p_3 * \tilde{k} * \tilde{k}\right) \\
&\quad + (q_0 * 1 * \mathbf{p} \cdot \mathbf{I} + p_0 * 1 * \mathbf{q} \cdot \mathbf{I}) \\
&\quad + (q_1 p_2 * \tilde{i} * \tilde{j} + q_2 p_1 * \tilde{j} * \tilde{i} \\
&\quad + q_2 p_3 * \tilde{j} * \tilde{k} + q_3 p_2 * \tilde{k} * \tilde{j} \\
&\quad + q_1 p_3 * \tilde{i} * \tilde{k} + q_3 p_1 * \tilde{k} * \tilde{i}) \\
&= (q_0 p_0 \;-\; \mathbf{q} \cdot \mathbf{p}) * 1 \;+\; (q_0 \mathbf{p} \;+\; p_0 \mathbf{q} \;+\; \mathbf{q} \times \mathbf{p}) \cdot \mathbf{I} .
\end{aligned} \tag{A.12}$$

Note that in Eq. (A.12) $\tilde{p}$ indicates a quaternion, and $\mathbf{p}$ the corresponding quaternion vector. The same style convention is also used for the pair $(\tilde{q}/\mathbf{q})$.

A.3.2 Quaternions and Rotation Matrices

Hypothesis The relationship between quaternions and rotation matrices is given by Eq. (4.14),

$$\tilde{x}' = \tilde{q} \circ \tilde{x} \circ \tilde{q}^{-1} = \begin{pmatrix} 0 \\ \mathbf{R} \cdot \mathbf{x} \end{pmatrix} . \tag{A.13}$$

where $\tilde{q}$ is a unit quaternions

Proof For this proof, the following basic trigonometric relationships will be needed:

$$\sin^2(\alpha) + \cos^2(\alpha) = 1 , \tag{A.14}$$

$$\sin(2\alpha) = 2 * \sin(\alpha) * \cos(\alpha)\,, \tag{A.15}$$

$$\cos(2\alpha) = \cos^2(\alpha) - \sin^2(\alpha)\,, \text{ and} \tag{A.16}$$

$$\mathbf{a} \times \mathbf{b} \times \mathbf{c} = \mathbf{b} * (\mathbf{a} \cdot \mathbf{c}) - \mathbf{c} * (\mathbf{a} \cdot \mathbf{b})\ . \tag{A.17}$$

Equations (A.14) and (A.16) also imply

$$1 - \cos(\alpha) = \left(\sin^2(\frac{\alpha}{2}) + \cos^2(\frac{\alpha}{2})\right) - \left(\cos^2(\frac{\alpha}{2}) - \sin^2(\frac{\alpha}{2})\right) = 2 * \sin^2(\frac{\alpha}{2}). \tag{A.18}$$

The first part of Eq. (A.13) is

$$\tilde{\boldsymbol{q}} \circ \tilde{\boldsymbol{x}} = \begin{pmatrix} q_0 \\ \mathbf{q} \end{pmatrix} \circ \begin{pmatrix} 0 \\ \mathbf{x} \end{pmatrix} = (-\mathbf{q} \cdot \mathbf{x}) + (q_0 * \mathbf{x} + \mathbf{q} \times \mathbf{x}) \cdot \mathbf{I}\,. \tag{A.19}$$

Inserting this into Eq. (A.13) leads to

$$\tilde{\boldsymbol{q}} \circ \tilde{\boldsymbol{x}} \circ \tilde{\boldsymbol{q}}^{-1} = \begin{pmatrix} -\mathbf{q} \cdot \mathbf{x} \\ q_0 * \mathbf{x} + \mathbf{q} \times \mathbf{x} \end{pmatrix} \circ \begin{pmatrix} q_0 \\ -\mathbf{q} \end{pmatrix}. \tag{A.20}$$

The scalar part of Eq. (A.20) is

$$- q_0 * \mathbf{q} \cdot \mathbf{x} + q_0 * \mathbf{q} \cdot \mathbf{x} + (\mathbf{q} \times \mathbf{x}) \cdot \mathbf{q} = 0\,. \tag{A.21}$$

And the vector part of Eq. (A.20) is

$$\begin{aligned} &(\mathbf{q} \cdot \mathbf{x}) * \mathbf{q} + q_0^2 * \mathbf{x} + q_0 * \mathbf{q} \times \mathbf{x} - q_0 * \mathbf{x} \times \mathbf{q} - (\mathbf{q} \times \mathbf{x} \times \mathbf{q}) = \\ &(\mathbf{q} \cdot \mathbf{x}) * \mathbf{q} + q_0^2 * \mathbf{x} + 2q_0 * \mathbf{q} \times \mathbf{x} - \left(\mathbf{x} * \mathbf{q}^2 - \mathbf{q} * (\mathbf{q} \cdot \mathbf{x})\right)\,. \end{aligned} \tag{A.22}$$

Since $q_0 = \cos(\frac{\theta}{2})$ and $\mathbf{q} = \mathbf{n} * \sin(\frac{\theta}{2})$, and using Eq. (A.18), Eq. (A.22) can be written as

$$\begin{aligned} &2 * \sin^2(\frac{\theta}{2}) * \mathbf{n} * (\mathbf{n} \cdot \mathbf{x}) + \mathbf{x} * \left(\cos^2(\frac{\theta}{2}) - \sin^2(\frac{\theta}{2})\right) + 2 * \sin(\frac{\theta}{2}) * \cos(\frac{\theta}{2}) * \mathbf{n} \times \mathbf{x} \\ &= (1 - \cos(\theta)) * \mathbf{n}(\mathbf{n} \cdot \mathbf{x}) + \mathbf{x}\cos(\theta) + \sin(\theta) * \mathbf{n} \times \mathbf{x}\,. \end{aligned} \tag{A.23}$$

Since the right side of Eq. (A.23) equals the Rodrigues representation of a rotation [(Eq. (4.2)], and since the scalar part of the equation is zero [(Eq. (A.21)], this completes the proof of Eq. (A.13).

A.3.3 Space-Fixed Versus Body-Fixed Rotations

In a space-fixed, orthogonal coordinate system, the components of a space-fixed point $\mathbf{p}_{\mathrm{sf}}$ are given by the projection of the point onto the coordinate axes. Assuming that vectors are by default column vectors

$$\mathbf{p} = \begin{pmatrix} p_x \\ p_y \\ p_z \end{pmatrix}_{\mathrm{sf}} \tag{A.24}$$

then i^{th} component of the vector can be obtained by

$$p_{sf,i} = \mathbf{p}^T \cdot \mathbf{e}_\mathrm{i} \,. \tag{A.25}$$

Hypothesis The orientation of the basis vectors of a rotated coordinate system is given by the columns of $\mathbf{R}$. But the vector components of a *space-fixed point* $\mathbf{p}$ transform with $\mathbf{R}^T$ into the rotated coordinate system.

Proof Rotation matrices are elements of the *special orthogonal group* SO_3, which means that

$$\mathbf{R}^T = \mathbf{R}^{-1}. \tag{A.26}$$

And since

$$(\mathbf{A} \cdot \mathbf{B})^T = \mathbf{B}^T \cdot \mathbf{A}^T \tag{A.27}$$

the coordinates of a space-fixed point $\mathbf{p}$, in a rotated coordinate system $\mathbf{e}'_\mathrm{i} = \mathbf{R} \cdot \mathbf{e}_\mathrm{i}$, are given by

$$\begin{aligned} p'_{bf,i} &= \mathbf{p}^T \cdot \mathbf{e}'_\mathrm{i} \\ &= \mathbf{p}^T \cdot \mathbf{R} \cdot \mathbf{e}_\mathrm{i} \\ &= (\mathbf{R}^T \cdot \mathbf{p})^T \cdot \mathbf{e}_\mathrm{i} \,. \end{aligned} \tag{A.28}$$

In summary

$$\mathbf{e}'_\mathrm{i} = \mathbf{R} \cdot \mathbf{e}_\mathrm{i} \,, \tag{A.29}$$

but

$$\begin{pmatrix} p'_x \\ p'_y \\ p'_z \end{pmatrix}_{\mathrm{bf}} = \mathbf{R}^T \cdot \begin{pmatrix} p_x \\ p_y \\ p_z \end{pmatrix}_{\mathrm{sf}} \,. \tag{A.30}$$

Or expressed in words: the coordinates (Eq. A.30) transform with the inverse of the transformation of the basis vectors (Eqs. 3.3 and A.29).

A.4 Clifford Algebra

The focus of this book has been to enable the user to work with recordings of position and orientation in three dimensions. While the equations for working with translations are predominantly intuitive, working with orientation can raise a number of questions:

- Why does the numerical cross product use the right-hand rule?
- Why do quaternions use the sine of *half* the angle, and not of the full angle?
- Where does the relation between imaginary numbers ($j^2 = -1$) and rotations ($e^{j\theta t} = \cos(\theta t) + j * \sin(\theta t)$) come from?

The answers to many of these questions can be obtained with *Clifford Algebras*. In the 1870s, William Kingdon Clifford sought to extend and unify Hamilton's quaternions with Hermann Grassmann's "extensive quantities" (Grassmann 1844) into a single algebra that Clifford called "geometric algebra". Clifford's geometric algebra has also been named after him in his honor, as "Clifford Algebra". Advantages of Clifford algebra include the following:

- The real numbers are a subalgebra of the Clifford algebra.
- Ordinary vector algebra is another subalgebra of the Clifford algebra.
- The complex numbers are another subalgebra of the Clifford algebra.
- Quaternions can be understood in terms of another subalgebra of Clifford algebra, namely the subalgebra containing just scalars and "bivectors" (see below). This is tremendously useful for describing rotations in three or more dimensions.

An introduction into Clifford algebras goes beyond the scope of this book. But a gentle introduction into the thinking underlying these algebras can be provided by showing how the scalar product and the wedge product (which is the generalization of the 3-D cross product) can be introduced in a coordinate-free way, independent of the dimensionality of the underlying vector space.

Also, note the excellent Python package `clifford`, with good corresponding documentation (http://clifford.readthedocs.io).

A.4.1 *Visualizing Scalars, Vectors, and More*

If the dimension is n, then there will be n independent basis vectors in the algebra. The algebra is generated by the *geometric products* (for definition see below) of these distinct basis vectors.

- The product of no basis vectors is the *scalar* 1. This is called a *grade 0* object and corresponds to a *point in space*.
- The product of one basis vector is the *vector* itself. This is called a *grade 1* object and corresponds to a *line segment*.
- The geometric product of two orthogonal distinct basis vectors is a *bivector*. This is called a *grade 2* object and corresponds to a *patch of surface*.
- A *trivector* is a *grade 3* object and corresponds to a *piece of space*.

Nonzero objects in the algebra can be made up by adding scalar multiples of any of the terms, drawn from any of the grades. Note that this violates the principle in vector analysis that a scalar cannot be added to a vector!

In practice, two common sorts of objects are vectors (all grade 1), and quaternions, which combine a scalar and a bivector.

A.4.2 Geometric Product

We now postulate that there exists a "geometric product" that can be used to multiply *any* element of the Clifford algebra with each other element. This geometric product shall have the following properties:

$$(AB)C = A(BC) = ABC \tag{A.31}$$

$$A(B + C) = AB + AC\,. \tag{A.32}$$

Equation (A.31) states that the geometric product is associative, and Eq. (A.32) that is distributive over addition.

Note that the geometric product is in general *not* commutative!

A.4.3 Dot Product and Wedge Product

With only these few assumptions, a "dot product" $P \cdot Q$ and a "wedge product" $P \wedge Q$ of two vectors P and Q can be introduced as follows:

$$P \cdot Q := \frac{PQ + QP}{2} \quad \text{where P and Q have } grade = 1 \tag{A.33}$$

$$P \wedge Q := \frac{PQ - QP}{2} \quad \text{where P and Q have } grade \leq 1\,. \tag{A.34}$$

The wedge product is sometimes called the "exterior product", and in three dimensions corresponds to the cross product.

A.4.4 A Practical Application of Clifford Algebra

With the few simple steps above, we have managed to define vectors, planes and dot and wedge products, for vector spaces with arbitrary dimensions, and without introducing a coordinate system!

Continuing with this approach, many physical laws can be formulated in a coordinate-free manner, thereby often massively simplifying the formulation, and clarifying the transformation properties. Take, for example, the famous Maxwell equations. In the original paper, Maxwell wrote his equations in component form, and as a consequence, there were 20 equations in his publication in 1865. Using vector notation, Heaviside in 1884 reduces these to the usually known four equations

$$\nabla \cdot \mathbf{E} = \frac{\rho}{\varepsilon_0} \qquad \text{Gauss's law}$$

$$\nabla \cdot \mathbf{B} = 0 \qquad \text{Gauss's law for magnetism}$$

$$\nabla \times \mathbf{E} = -\frac{\partial \mathbf{B}}{\partial t} \qquad \text{Faraday's law}$$

$$\nabla \times \mathbf{B} = \mu_0 \mathbf{J} + \mu_0 \varepsilon_0 \frac{\partial \mathbf{E}}{\partial t} \qquad \text{AmpèreMaxwell law}$$

But using geometric algebra, these can be reduced to the single equation[1]

$$\nabla F = \mu_0 c J \ , \tag{A.35}$$

where F is the multi-vector describing the electromagnetic field, and J is the current multi-vector describing resting and moving charges. Using this notation, properties such as the relativistic invariance of this equation with respect to the observer's time and space coordinates are much easier to prove than in the vector notation.

A.4.5 Relationship to Rotations

[2]In all dimensions, rotations are fully described by the planes of the rotation and their associated angles, so it is useful to be able to determine them, or at least find ways to describe them mathematically. This section also explains why the length of the vector part of quaternions is only *half* the angle of the corresponding rotation. Mathematical proofs are left away here, in order to facilitate the introduction to the mathematically more complex arena of Clifford algebras.

[1]https://en.wikipedia.org/wiki/Mathematical_descriptions_of_the_electromagnetic_field.

[2]This part has been adopted from https://en.wikipedia.org/wiki/Plane_of_rotation.

Reflections

Every simple rotation can be generated by two reflections (Fig. A.4). Reflections can be specified in n dimensions by giving an $(n-1)$-dimensional subspace to reflect in: a two-dimensional reflection is a reflection on a line, a three-dimensional reflection is a reflection on a plane, and so on. But this becomes increasingly difficult to apply in higher dimensions, so it is better to use vectors instead, as follows.

A reflection in n dimensions is specified by a vector perpendicular to the $(n-1)$-dimensional subspace. To generate simple rotations only reflections that fix the origin are needed, so the vector does not have a position, just a direction. It also does not matter which way it is facing: it can be replaced with its negative without changing the result. Similarly, unit vectors can be used to simplify the calculations.

So the reflection on an $(n-1)$-dimensional space is given by the unit vector perpendicular to it, $\mathbf{m}$, with:

$$\mathbf{x}' = -\mathbf{mxm}, \tag{A.36}$$

where the product is the geometric product from Clifford algebra.

If $\mathbf{x}'$ is reflected in another, distinct, $(n-1)$-dimensional space, described by a unit vector $\mathbf{n}$ perpendicular to it, the result is

$$\mathbf{x}'' = -\mathbf{nx}'\mathbf{n} = -\mathbf{n}(-\mathbf{mxm})\mathbf{n} = \mathbf{nmxmn}\,. \tag{A.37}$$

This is a simple rotation in n dimensions, through twice the angle between the subspaces, which is also the angle between the vectors $\mathbf{m}$ and $\mathbf{n}$. It can be checked using geometric algebra that this is a rotation, and that it rotates all vectors as expected.

The quantity $\mathbf{mn}$ is a called a *rotor*, and $\mathbf{nm}$ is its inverse as

$$(\mathbf{mn})(\mathbf{nm}) = \mathbf{mnnm} = \mathbf{mm} = 1\,. \tag{A.38}$$

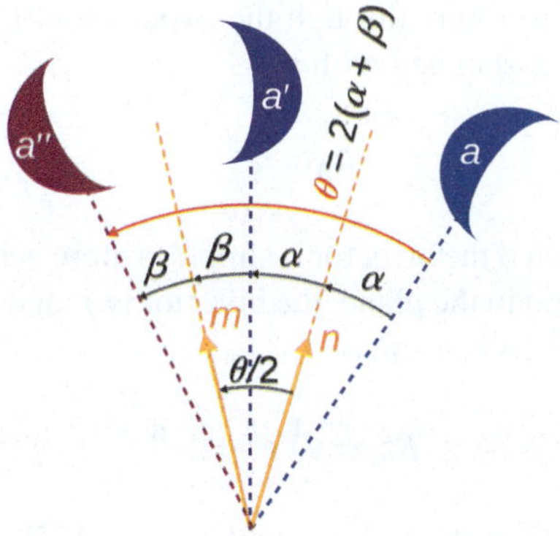

Fig. A.4 Two different reflections in 2-D, one about **m** and one about **n**, generating a rotation. (by Maschen)

So the rotation can be written as

$$\mathbf{x}'' = R\mathbf{x}R^{-1}, \tag{A.39}$$

where $R = \mathbf{mn}$ is the rotor.

The plane of rotation is the plane containing **m** and **n**, which must be distinct otherwise the reflections are the same and no rotation takes place. As either vector can be replaced by its negative the angle between them can always be acute, or at most $\pi/2$. The rotation is through twice the angle between the vectors, up to π or a half-turn. The sense of the rotation is to rotate from **m** toward **n**: the geometric product is not commutative so the product **nm** is the inverse rotation, with sense from **n** to **m**.

Conversely, all simple rotations can be generated this way, with two reflections, by two unit vectors in the plane of rotation separated by half the desired angle of rotation. These can be composed to produce more general rotations, using up to n reflections if the dimension n is even, $n-2$ if n is odd, by choosing pairs of reflections given by two vectors in each plane of rotation.

Bivectors

Bivectors are quantities from Clifford algebra, which generalize the idea of (line-)vectors. As vectors are to lines, so are bivectors to planes. So every plane (in any dimension) can be associated with a bivector, and every simple bivector is associated with a plane. This makes them a good fit for describing planes of rotation. They are behind the rotation by $e^{i\theta}$ in the complex plane (Sect. 3.2.2), and the quaternion vectors in three dimensions.

Every rotation has a simple bivector associated with it. This is parallel to the plane of the rotation and has a magnitude equal to the angle of rotation in the plane. These bivectors are summed to produce a single, generally non-simple, bivector for the whole rotation. This can generate a rotor through the exponential map, which can be used to rotate an object.

Bivectors are related to rotors through the exponential map. In particular, given any bivector B the rotor associated with it is

$$R_{\mathbf{B}} = e^{\frac{\mathbf{B}}{2}}. \tag{A.40}$$

This is a simple rotation if the bivector is simple, a more general rotation otherwise. For example, for a rotation in the plane, the bivector is j, and the rotor corresponding to a rotation by θ is $e^{j\theta/2}$. When squared,

$$R_{\mathbf{B}}^2 = e^{\frac{\mathbf{B}}{2}} e^{\frac{\mathbf{B}}{2}} = e^{\mathbf{B}} \tag{A.41}$$

it gives a rotor that rotates through twice the angle. If **B** is simple then this is the same rotation as is generated by two reflections, as the product **mn** gives a rotation through twice the angle between the vectors. (Compare that to Eq. (4.11), which shows the exponential of a quaternion.) These can be equated,

$$\mathbf{mn} = e^{\mathbf{B}}, \tag{A.42}$$

from which it follows that the bivector associated with the plane of rotation containing **m** and **n** that rotates **m** to **n** is

$$\mathbf{B} = \log(\mathbf{mn}). \tag{A.43}$$

This is a simple bivector, associated with the simple rotation described. More general rotations in four or more dimensions are associated with sums of simple bivectors, one for each plane of rotation, calculated as above.

A.5 Spherical Statistics

In one dimension, (normal) distributions are commonly characterized by their mean value and their standard deviation. On circles and spheres, though, those parameters are harder to characterize. Take, for examples, two orientation measurements, the first one being 10°, and the second one 350°.

Calculating the mean value in the classical way, one obtains $mean = \frac{10+350}{2} = 180°$, while the correct value is 0° (see Fig. A.5)! (Note that the problem can only be shifted, but not solved by using the interval −180° to +180° instead of 0° to 360°.) Given this problem, which procedure can be used to find the correct average orientation for multiple points? The solution is remarkably simple: converting every angle (solid lines in Fig. A.5) to a point on the unit circle (red points in Fig. A.5), and taking the mean value of those points (black point in Fig. A.5) provides the

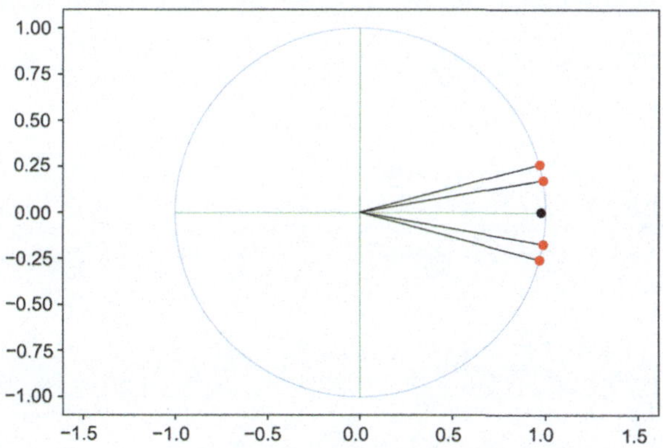

Fig. A.5 The *mean angle* of the red dots on a unit circle is given by the angle of the black dot. And their *dispersion* is characterized by the distance of the black dot from the unit circle

correct *mean angle*. And the distance of that point from the unit circle provides the "dispersion", a parameter that characterizes the spread of the angles. This method also generalizes trivially to three dimensions.

An application example for directional statistics is provided by (Leong and Carlile 1998). For more involved calculations of spherical statistics, e.g., comparison of two distributions, characterization of skewed distributions, etc., see the books (Mardia and Jupp 1999; Ley and Verdebout 2017).

Appendix B
Practical Applications: Denavit-Hartenberg Transformations

An elegant example of the application of homogeneous coordinates are the Denavit-Hartenberg transformations. In mechanical engineering, the "Denavit-Hartenberg parameters" (also called "DH parameters") are the four parameters associated with a particular convention for attaching reference frames to the links of robot manipulators.

In this convention, coordinate frames are attached to the joints between two links such that one transformation is associated with the joint, [Z], and the second is associated with the link [X]. The coordinate transformations along a serial robot consisting of n links form the kinematic equations of the robot,

$$\mathbf{T} = \mathbf{Z}_1 \cdot \mathbf{X}_1 \cdot \mathbf{Z}_2 \cdot \mathbf{X}_2 \ldots \cdot \mathbf{X}_{n-1} \cdot \mathbf{Z}_n \cdot \mathbf{X}_n, \tag{B.1}$$

where $\mathbf{T}$ is the transformation locating the end-link (see Fig. B.1).

So going "from the inside out", i.e. starting with the rotation-translation that does not affect any other joints: first the end-link is rotated about the x-axis, by an angle α. And it is translated along the x-axis by a distance r. (Note that the sequence in which this translation-rotation is executed has no consequence on the final position-orientation of the end-link.) This rotation-translation is described by the spatial transformation matrix

$$\mathbf{X}_i = \begin{bmatrix} 1 & 0 & 0 & r_i \\ 0 & \cos\alpha_i & -\sin\alpha_i & 0 \\ 0 & \sin\alpha_i & \cos\alpha_i & 0 \\ 0 & 0 & 0 & 1 \end{bmatrix}, \tag{B.2}$$

The second rotation-translation is about the original z-axis, by a distance d, and by an angle θ. That sequence is described by the spatial transformation matrix

T. Haslwanter, *3D Kinematics*, https://doi.org/10.1007/978-3-319-75277-8

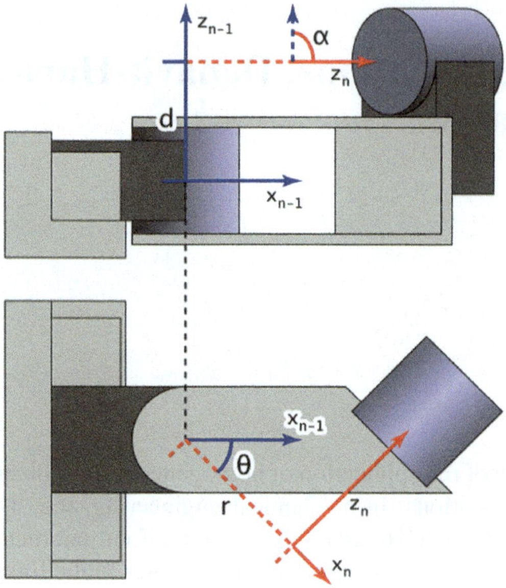

Fig. B.1 Illustrates the transformation parameters of a pair of reference frames laid out according to Denavit-Hartenberg convention. (Illustration by Ethan Tira-Thompson, from WikiMedia Commons)

$$\mathbf{Z}_i = \begin{bmatrix} \cos\theta_i & -\sin\theta_i & 0 & 0 \\ \sin\theta_i & \cos\theta_i & 0 & 0 \\ 0 & 0 & 1 & d_i \\ 0 & 0 & 0 & 1 \end{bmatrix}, \tag{B.3}$$

The combined state-matrix for one link

$$T_i = Z_i * X_i \tag{B.4}$$

depends on four parameters (d, θ, r, α).

The equations for practical implementations are available in the function `rotmat.dh` for numerical calculations, and in `rotmat.dh_s` if the symbolic equations are required. For example

```
from skinematics.rotmat import dh_s

T_final = dh_s('theta', 'd', 'r', 'alpha')
translation = T_final[:3,-1]
rotation = T_final[:3, :3]
print(translation)

>> Matrix([[r*cos(theta)], [r*sin(theta)], [d]])
```

shows that the location of the end-joint depends only on (r, d, θ). And its orientation is given by the rotation matrix `R_final`.

And a numeric example, with a robotic chain with three links, where the links have lengths of *30cm, 50cm, 20cm*, and the angles of rotations (θ, α) are *45, 30, 60* and *10, 20, 30* deg, respectively, and where there are no offsets from the rotation axes:

```
from skinematics.rotmat import dh

rs = [0.3, 0.5, 0.2] # in [m]
thetas = [45, 30, 60] # in [deg]
alphas = [10, 20, 30] # in [deg]
T = np.eye(4)        # start in the reference position/
    orientation

for r, theta, alpha in zip(rs, thetas, alphas):
    T = T @ dh(theta=theta, alpha=alpha, r=r)

translation = T_final[:3,-1]
rotation = T_final[:3, :3]
```

Appendix C
Python and Matlab Programs

One goal of this book is to provide the reader not only with explanations and examples for 3-D kinematics, but also with a list of programs allowing quick implementation of analytical procedures that are typically encountered while working with 3-D movement recordings.

Programming libraries are available for Python, *scikit-kinematics*, and Matlab, 3-D *Kinematics toolbox* (see Sect. 1.3). Most functions in the Matlab 3-D *Kinematics* toolbox have the same name, and use the same arguments, as the corresponding *scikit-kinematics* functions. The only exceptions are function names that would collide with existing Matlab functions, or functions which would have the same name in the unified Matlab namespace.

Functions are commonly written in such a way that they can be applied to a single point/vector/quaternion, or simultaneously to a whole data set. Data sets are thereby always stored such that one row corresponds to one data set (see the first code example on p. 135).

This section gives an overview of these functions and their application. For more detailed information, see the documentation on http://work.thaslwanter.at/skinematics/html/.

C.1 List of Programs

See Appendix Table C.1.

C.2 Vector Calculations

The module `vectors` in the package `scikit-kinematics` provides the functionalities shown below. Note that all the functions are vectorized, i.e., they work with matrix inputs, allowing simultaneous application of the function to each input line.

T. Haslwanter, *3D Kinematics*, https://doi.org/10.1007/978-3-319-75277-8

Table C.1 Python and Matlab functions provided with this book

Group	Python	Matlab
quat	Quaternion (class)	@quat(class)
	calc_angvel	calc_angvel
	calc_quat	calc_quat
	convert	quat_convert
	deg2quat	deg2quat
	q_conj	q_conj
	q_inv	q_inv
	q_mult	q_mult
	q_scalar	q_scalar
	q_vector	q_vector
	quat2deg	quat2deg
	quat2seq	quat2seq
	unit_q	unit_q
rotmat	R	R
	dh	dh
	dh_s	dh_s
	R	R
	R_s	R_s
	seq2quat	seq2quat
	sequence	sequence
imus	IMU_Base (class)	—
	analytical	analyze_imus
	kalman	imu_Kalman
	Mahony	@MahonyAHRS(class)
	Madgwick	@MadgwickAHRS(class)
		imu_Madgwick, or imu_Mahony
markers	analyze_3Dmarkers	analyze_3Dmarkers
	find_trajectory	find_trajectory
vector	angle	vector_angle
	GramSchmidt	GramSchmidt
	normalize	normalize
	project	project_vector
	plane_orientation	plane_orientation
	q_shortest_rotation	q_shortest_rotation
	rotate_vector	rotate_vector
	target2orient	target2orient
view	orientation	view_orientation
	ts	view_ts
utility	— (in numpy)	copysign
	— (in numpy)	deg2rad
	(in Class IMU)	get_XSens
	— (in numpy)	rad2deg
	— (in scipy)	savgol
	— (in numpy)	toRow

For example, normalizing a single vector one obtains

```
import skinematics as skin
v = [1,2,3]
skin.vector.normalize(v)
>> array([ 0.267, 0.535, 0.802])
```

But this also works for arrays:

```
import numpy as np
v = np.random.randn(1000,3)
skin.vector.normalize(v)
>> array([[-0.628, -0.663, 0.408],
          [ 0.664, 0.625, -0.41 ],
          [-0.547, -0.038, -0.837],
          ...,
          [-0.205, 0.061, -0.977],
          [-0.559, -0.734, -0.386],
          [ 0.105, 0.658, -0.746]])
```

Below is a list of the functions in the module `vectors`, and the preamble required to execute the examples (note that in Python function names are commonly preceded by the module they are in):

```
# Import the required packages and functions
import numpy as np
from skinematics import vector

# Input data
v0 = [0, 0, 0]
v1 = [1, 0, 0]
v2 = [1, 1, 0]
```

vector.angle Angle between two vectors.

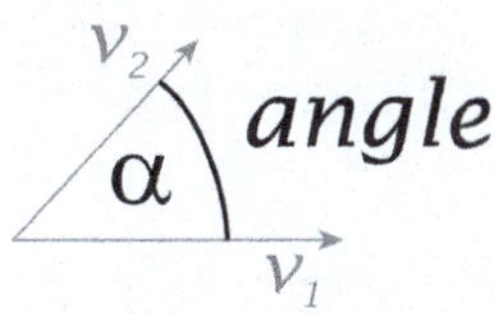


```
angle = vector.angle(v1, v2)
print('The angle between v1 and v2 is {0:4.1f}
    degree\n'.format(np.rad2deg(angle)))
>> The angle between v1 and v2 is 45.0 degree
```

vector.GramSchmidt Gram–Schmidt orthogonalization of three points $\mathbf{p_0}$, $\mathbf{p_1}$, $\mathbf{p_2}$. In the figure below, $\mathbf{a}_i$, $i = x, y, z$ are the unit vectors of an orthogonal coordinate system (more about that in Chap. 6).

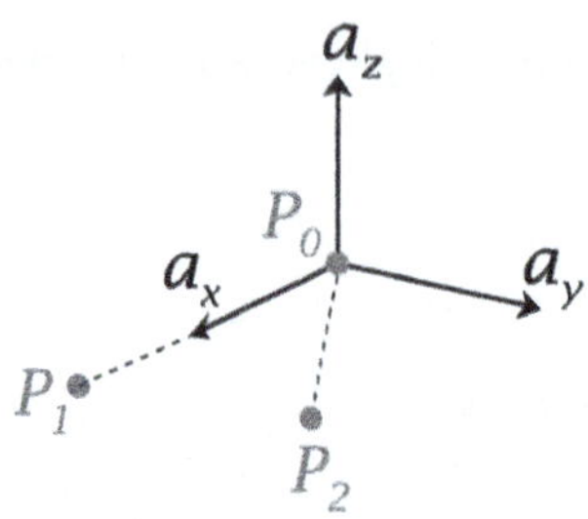


```
# Gram-Schmidt orthogonalization
gs = vector.GramSchmidt(v0, v1, v2)
print('The Gram-Schmidt orthogonalization of the points v0,
     v1, and v2 is {0}\n'.format(np.reshape(gs, (3,3))))
>> The Gram-Schmidt orthogonalization of the points p0, p1,
     and p2 is
[[ 1.  0.  0.]
 [ 0.  1.  0.]
 [ 0.  0.  1.]]
```

vector.normalize Normalization of a vector **v**.

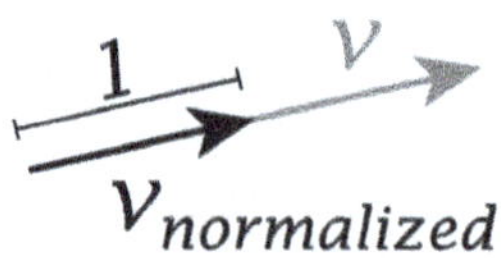


```
# Vector normalization
v2_normalized = vector.normalize(v2)
print('v2 normalized is {0}\n'.format (v2_normalized))
>> v2 normalized is [ 0.70710678 0.70710678 0.          ]
```

vector.project Projection of a vector $\mathbf{v_1}$ onto a vector $\mathbf{v_2}$.

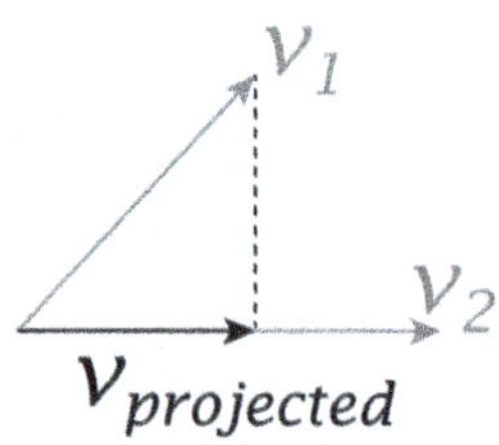


```
# Projection
projected = vector.project(v1, v2)
print('The projection of v1 onto v2 is {0}\n'.format
(projected))
>> The projection of v1 onto v2 is [ 0.5 0.5 0. ]
```

vector.plane_orientation Vector **n** perpendicular to the plane defined by the three points $\mathbf{p_0}$, $\mathbf{p_1}$, $\mathbf{p_2}$.

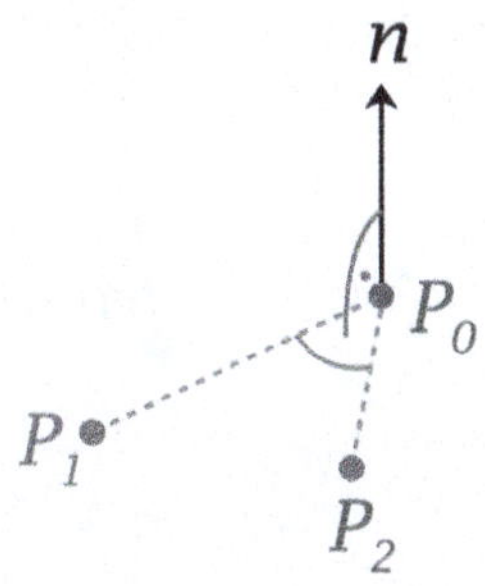


```
# Plane orientation
n = vector.plane_orientation(v0, v1, v2)
print('The plane spanned by v0, v1, and v2 is
    orthogonal to {0}\n'.format(n))
>> The plane spanned by p0, p1, and p2 is
    orthogonal to [ 0. 0. 1.]
```

vector.q_shortest_rotation Quaternion vector **q**, indicating the shortest rotation that rotates $\mathbf{v_1}$ into $\mathbf{v_2}$ (more about that in Chap. 4).

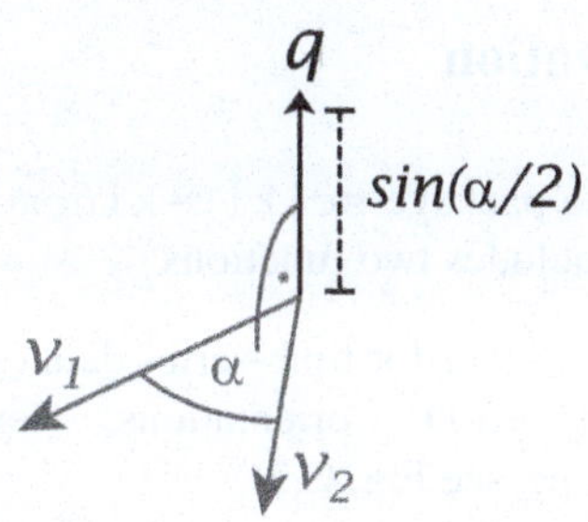


```
# Shortest rotation
q_shortest = vector.q_shortest_rotation(v1, v2)
print('The shortest rotation that brings v1 in alignment
    with v2 is described by the quaternion {0}\n'.format
(q_shortest))
>> The shortest rotation that brings v1 in alignment with
    v2 is described by the quaternion
[ 0.         0.          0.38268343]
```

vector.rotate_vector Rotation of a vector, with a quaternion (more about that in Chap. 4).

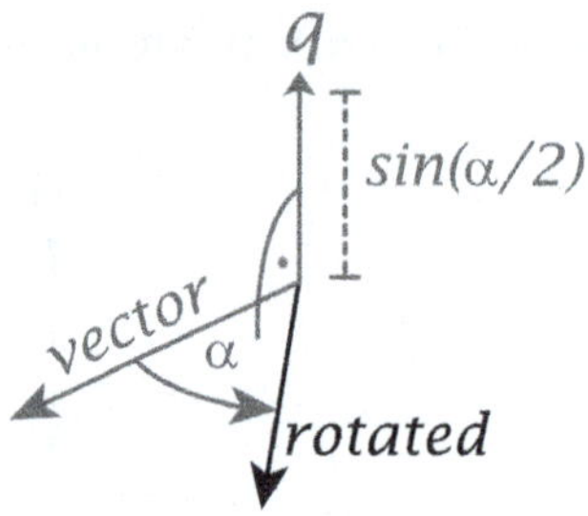


```
# Rotation of a vector by a quaternion
q = [0, 0.1, 0]
rotated = vector.rotate_vector(v1, q)
print('v1 rotated by {0} is: {1}'.format(q, rotated))
>> v1 rotated by [0, 0.1, 0] is:
[ 0.98       0.         -0.19899749]
```

Code: `C1_examples_vectors.py`: Contains the examples of working with vectors presented above.

C.3 Data Visualization

The module `view` in the package `scikit-kinematics` provides a viewer for 3-D data. This module includes two functions:

- An interactive, simple viewer for time-series data (`view.ts`, see Fig. C.1).
- An animation of 3-D orientations, expressed as quaternions (`view.orientation`, see Fig. C.2).

For `view.ts` the following options are available, in addition to the (obvious) GUI interactions:

Keyboard shortcuts:

f forward (+ 1/2 frame)
n next (+ 1 frame)
b back (−1/2 frame)
p previous (−1 frame)
z zoom (x-frame = 10% of total length)
a all (adjust x- and y-limits)
x exit

Optimized y-scale: Often one wants to see data symmetrically about the zero-axis. To facilitate this display, adjusting the Upper Limit automatically sets the lower limit to the corresponding negative value.

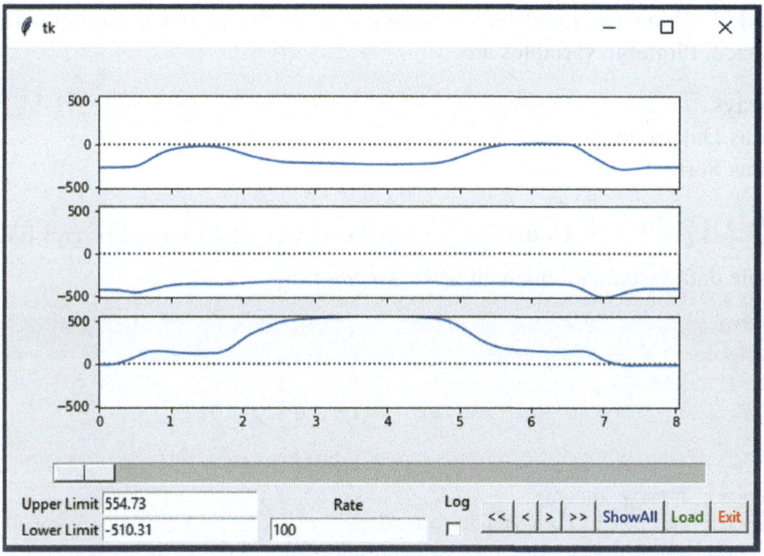

Fig. C.1 Time-series viewer `view.tx` from *scikit-kinematics*. The data show the (x, y, z)-component of the lower arm position during a drinking movement

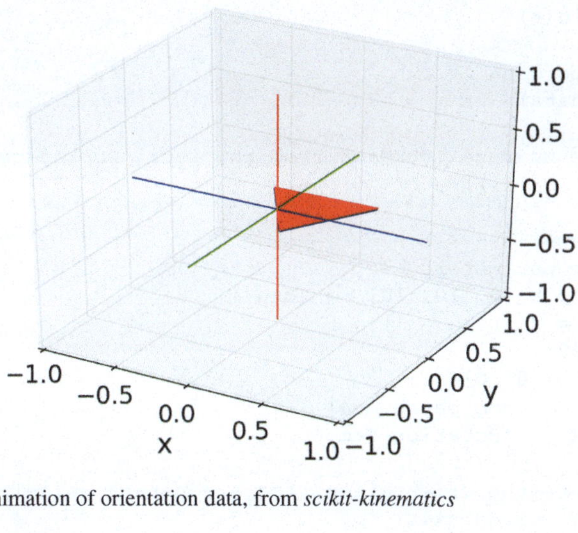

Fig. C.2 3-D animation of orientation data, from *scikit-kinematics*

Logging: When Log is activated, right-mouse clicks are indicated with vertical bars, and the corresponding x-values are stored in the users home-directory, in the file [varName].log. Since the name of the first value is unknown the first events are stored into data.log.

Load: Pushing the Load button shows you all the plottable variables in your namespace. Plottable variables are:

- ndarrays
- Pandas DataFrames
- Pandas Series

python **Code**: `C1_examples_visualization.py`: Example of simple data visualizations with *scikit-kinematics*.

Listing : C1_examples_visualization.py

```
"""Visualizing 3-D data. """
# author: Thomas Haslwanter, date: July-2017

# Import the required packages and functions
import numpy as np
import matplotlib.pyplot as plt
from skinematics import view, quat

# 2D Viewer -----------------
data = np.random.randn(100,3)
t = np.arange(0,2*np.pi,0.1)
x = np.sin(t)

# Show the data
view.ts(data)

# Let the user select data from the local workspace
view.ts(locals())

# 3-D Viewer ----------------
# Set the parameters
omega = np.r_[0, 10, 10]   # [deg/s]
duration = 2
rate = 100
q0 = [1, 0, 0, 0]
out_file = 'demo_patch.mp4'
title_text = 'Rotation Demo'

# Calculate the orientation, for a constant velocity rotation
# about a tilted axis
dt = 1./rate
num_rep = duration*rate
omegas = np.tile(omega, [num_rep, 1])
q = quat.calc_quat(omegas, q0, rate, 'sf')

#orientation(q)
view.orientation(q, out_file, 'Well done!')
```

C.4 Rotation Matrices

C.4.1 Functions in scikit-kinematics.rotmat

The module `rotmat` in the package `scikit-kinematics` provides the following functionality:

rotmat.R Defines a 3-D rotation matrix for a rotation about a coordinate axis.
rotmat.convert Converts a rotation matrix to the corresponding quaternion (more about quaternions in Chap. 4).
rotmat.sequence Converts a rotation matrix to the corresponding nautical angles, Euler angles, etc.
rotmat.seq2quat Calculation of quaternions from Euler, Fick/nautical, Helmholtz angles.
rotmat.R_s Symbolic matrix for rotation about a coordinate axis.

python™ **Code**: `C3_examples_rotmat.py`: Example of working with rotation matrices.

Listing : C3_examples_rotmat.py

```
"""Working with rotation matrices """

# author: Thomas Haslwanter, date: Dec-2017

# Import the required packages and functions
import numpy as np
from skinematics import rotmat
from pprint import pprint

# Since I use R and R_s repeatedly, I import them directly
from skinematics.rotmat import R, R_s

# Rotation about the x-axis, by 30 deg
Rx_30 = R(axis='x', angle=30)

print('Rotation matrix describing a rotation about the x-axis
     by 30 deg:')
print(Rx_30)

# Find the rotation matrix for the nautical sequence
# "pprint" is required for a nicer, matrix-shaped display
R_nautical = R_s('z', 'theta') * R_s('y', 'phi') * R_s('x', '
    psi')

print('\nNautical sequence:')
pprint(R_nautical)
```

```
# Rotation matrix for Euler sequence, for given angles
alpha, beta, gamma = 45, 30, 20 # in [deg]
R = R('z', gamma) @ R('x', beta) @ R('z', alpha)

print('\nRotation matrix, for "Euler-angles" of {0}, {1}, {2}
     deg:'.format(alpha, beta, gamma))
print(R)

# Corresponding nautical sequence:
nautical = rotmat.sequence(R, to='nautical')

# ... and just to check
euler = rotmat.sequence(R, to='Euler')

print('\nNautical sequence: {0}'.format(nautical))
print('Euler sequence: {0}'.format(np.rad2deg(euler)))

'''
Output
------
Rotation matrix describing a rotation about the x-axis by 30
    deg:
[[ 1.         0.          0.        ]
 [ 0.         0.8660254 -0.5       ]
 [ 0.         0.5         0.8660254]]

Nautical sequence:
Matrix([
[cos(phi)*cos(theta), sin(phi)*sin(psi)*cos(theta) - sin
(theta)*cos(psi), sin(phi)*cos(psi)*cos(theta) + sin(psi)*sin
    (theta)],
[sin(theta)*cos(phi), sin(phi)*sin(psi)*sin(theta) + cos(psi)
    *cos(theta), sin(phi)*sin(theta)*cos(psi) - sin(psi)*
    cos(theta)],
[         -sin(phi),                       sin(psi)*cos(phi),
                                    cos(phi)*cos(psi)]])

Rotation matrix, for "Euler-angles" of 45, 30, 20 deg:
[[ 0.45501932 -0.87390673 0.17101007]
 [ 0.81728662 0.3335971 -0.46984631]
 [ 0.35355339 0.35355339 0.8660254 ]]

Nautical sequence: [ 1.06279023 -0.36136712 0.38759669]
Euler sequence: [ 45. 30. 20.]
'''
```

C.4.2 Symbolic Computations

Calculating 3-D rotation matrices by hand is not only tedious but also prone to errors. Luckily, in Python as well as in Matlab tools are available which allow symbolic computation, in order to produce, e.g., Eq. (3.23) rather easily. In Python, it is the package `sympy` (http://www.sympy.org/en/index.html), and in Matlab the *Symbolic Computation Toolbox*.

In Python, only two sub-steps are necessary:

1. Symbolic variables have to be declared explicitly, by defining them as objects from the class `sympy.Symbol`.
2. For working with symbolic matrices, the matrices have to be generated with `sympy.Matrix`.

The module `rotmat` in the package `scikit-kinematics` takes care of all these steps and makes it easy to generate and manipulate symbolic matrices for rotations about the three cardinal axes. For example, the following listing shows how the formula in Eq. (3.23) can be generated computationally:

```
from skinematics import rotmat
from pprint import pprint

Rx = rotmat.R_s('x', 'psi')
Ry = rotmat.R_s('y', 'phi')
Rz = rotmat.R_s('z', 'theta')
R_nautical = Rz * Ry * Rx

pprint(R_nautical)
```

The package `pprint` is only used to print the matrices in a more appealing and easier to read way.

`sympy` also allows substitution of symbolic variables with the corresponding values. For example, to obtain the matrix `R_nautical` for `psi = 0`, one can simply type

```
R_nautical.subs('psi', 0)
```

python™ **Code**: `C2_symbolic_computation.py`: Explicit implementation of how Eq. (3.23) can be determined computationally, to show how symbolic computations can be implemented.

C.5 Quaternions

The module `quat` in the package `scikit-kinematics` provides the following functionality for working with orientations (functions relating to angular velocity are described in Chap. 5):

Working with Quaternions

quat.q_conj Conjugate quaternion.
quat.q_inv Quaternion inversion.
quat.q_mult Quaternion multiplication.
quat.q_scalar Extract the scalar part from a quaternion.
quat.q_vector Extract the vector part from a quaternion.
quat.unit_q Extend a quaternion vector to a unit quaternion.

Conversion routines

quat.convert Convert quaternion to corresponding rotation matrix or Gibbs vector.
quat.deg2quat Convert a number or axis angles to quaternion vectors.
quat.quat2seq Convert quaternions to corresponding rotation angles.
quat.scale2deg Convert quaternion to corresponding axis angle.

In addition, the class `Quaternion` allows object-oriented programming with quaternion objects, with multiplication, division, and inverse. A `Quaternion` can be created from vectors, rotation matrices, or from nautical angles, Helmholtz angles, or Euler angles. The class provides

- operator overloading for *mult*, *div*, and *inv*, with the corresponding quaternion functions.
- indexing, and
- access to the data, in the attribute `values`.

python **Code**: `C4_examples_quat.py`: Example of working with quaternions.

Listing : C4_examples_quat.py

```
"""Working with quaternions """

# author: Thomas Haslwanter, date: Dec-2017

# Import the required functions
import numpy as np
from skinematics import quat
from pprint import pprint
import matplotlib.pyplot as plt

# Just the magnitude
q_size = quat.deg2quat(10)
print('10 deg converted to quaternions is
{0:5.3f}\n'.format(q_size))

# Input quaternion vector
alpha = [10, 20]
print('Input angles: {0}'.format(alpha))
```

```
alpha_rad = np.deg2rad(alpha)
q_vec = np.array([[0, np.sin(alpha_rad[0]/2), 0],
                  [0, 0, np.sin(alpha_rad[1]/2)]])
print('Input:')
pprint(q_vec)


# Unit quaternion
q_unit = quat.unit_q(q_vec)
print('\nUnit quaternions:')
pprint(q_unit)

# Also add a non-unit quaternion
q_non_unit = np.r_[1, 0, np.sin(alpha_rad[0]/2), 0]
q_data = np.vstack((q_unit, q_non_unit))
print('\nGeneral quaternions:')
pprint(q_data)

# Inversion
q_inverted = quat.q_inv(q_data)
print('\nInverted:')
pprint(q_inverted)

# Conjugation
q_conj = quat.q_conj(q_data)
print('\nConjugated:')
pprint(q_conj)

# Multiplication
q_multiplied = quat.q_mult(q_data, q_data)
print('\nMultiplied:')
pprint(q_multiplied)

# Scalar and vector part
q_scalar = quat.q_scalar(q_data)
q_vector = quat.q_vector(q_data)

print('\nScalar part:')
pprint(q_scalar)
print('Vector part:')
pprint(q_vector)

# Convert to axis angle
q_axisangle = quat.quat2deg(q_unit)
print('\nAxis angle:')
pprint(q_axisangle)

# Conversion to a rotation matrix
rotmats = quat.convert(q_unit)
print('\nFirst rotation matrix')
```

```
pprint(rotmats[0].reshape(3,3))


# Working with Quaternion objects
# --------------------------------
data  = np.array([ [0,0,0.1], [0, 0.2, 0 ] ])
data2 = np.array([ [0,0,0.1], [0, 0, 0.1] ])

eye = quat.Quaternion(data)
head = quat.Quaternion(data2)

# Quaternion multiplication, ...
gaze = head * eye

# ..., division, ...
head = gaze/eye
# or, equivalently
head = gaze * eye.inv()

# ..., slicing, ...
print(head[0])

# ... and access to the data
head_values = head.values
print(type(head.values))
'''

Output
------
10 deg converted to quaternions is 0.087

Input angles: [10, 20]
Input:
array([[ 0.        ,  0.08715574, 0.        ],
       [ 0.        ,  0.        ,  0.17364818]])

Unit quaternions:
array([[ 0.9961947 , 0.         ,  0.08715574, 0.         ],
       [ 0.98480775, 0.         ,  0.        ,  0.17364818]])

General quaternions:
array([[ 0.9961947 , 0.         ,  0.08715574, 0.         ],
       [ 0.98480775, 0.         ,  0.        ,  0.17364818],
       [ 1.        ,  0.        ,  0.08715574, 0.         ]])

Inverted:
array([[ 0.9961947 , -0.        , -0.08715574, -0.        ],
       [ 0.98480775, -0.        , -0.        , -0.17364818],
       [ 0.99246114, -0.        , -0.08649869, -0.        ]])

Conjugated:
```

```
array([[ 0.9961947 , -0.        , -0.08715574, -0.        ],
       [ 0.98480775, -0.        , -0.        ,  -0.17364818],
       [ 1.        , -0.        , -0.08715574, -0.        ]])

Multiplied:
array([[ 0.98480775, 0.        , 0.17364818, 0.        ],
       [ 0.93969262, 0.        , 0.        , 0.34202014],
       [ 0.99240388, 0.        , 0.17431149, 0.        ]])

Scalar part:
array([ 0.9961947 , 0.98480775, 1.        ])
Vector part:
array([[ 0.        , 0.08715574, 0.        ],
       [ 0.        , 0.        , 0.17364818],
       [ 0.        , 0.08715574, 0.        ]])

Axis angle:
array([[ 0., 10., 0.],
       [ 0.,  0., 20.]])

First rotation matrix
array([[ 0.98480775, 0.        , 0.17364818],
       [ 0.        , 1.        , 0.        ],
       [ -0.17364818, 0.        , 0.98480775]])

Quaternion [[ 0.99498744 0.          0.          0.1        ]]

<class 'numpy.ndarray'>
'''
```

C.6 Angular Velocity

The module quat in the package scikit-kinematics provides also the following functionality for working with angular velocities (Figs. C.3 and C.4)

quat.calc_angvel Calculates the angular velocity in space from quaternions.
quat.calc_quat Calculates the orientation from a starting orientation and angular velocity.

python™ **Code**: C5_examples_vel.py: Example of working with quaternions.

Listing : C5_examples_vel.py

```
"""Working with angular velocities """

# author: Thomas Haslwanter, date: Sept-2017
```

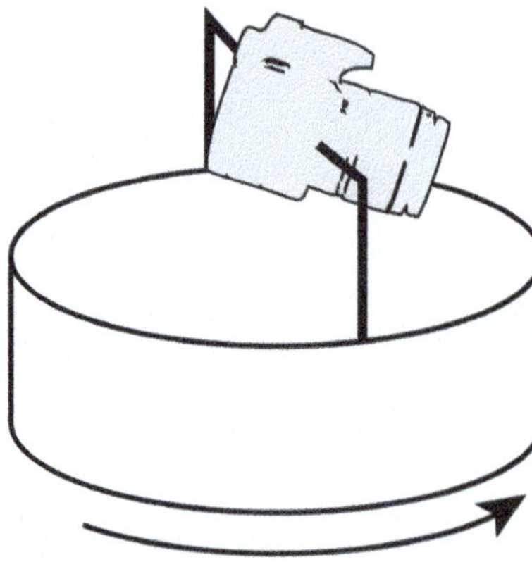

Fig. C.3 Rotating camera, looking downward

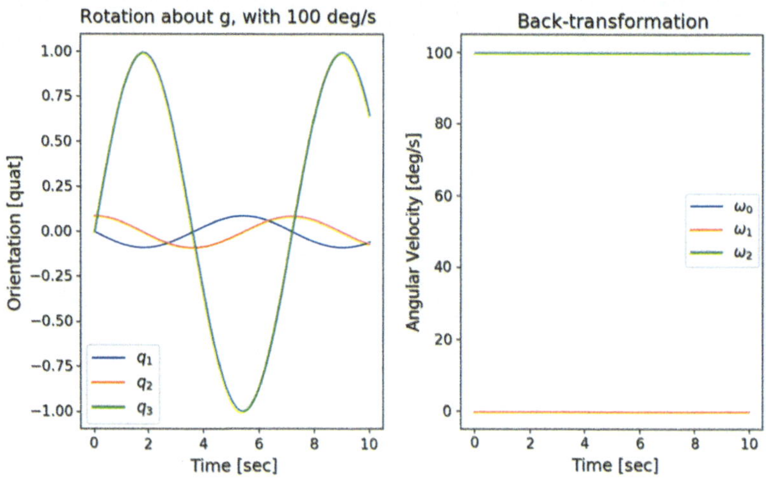

Fig. C.4 Quaternion describing the orientation of a camera pointing 10° down, on a platform rotating with $100° s^{-1}$ about an earth vertical axis. Note the dynamic change of the quaternion in the left figure, despite the fact that the orientation of the camera relative to the platform is fixed, and that the platform rotates exactly about the vertical axis!

```
# Import the required functions
import numpy as np
from skinematics import quat
import matplotlib.pyplot as plt

# Camera Looking 10 deg down, and rotating with 100 deg/s
    about an
# earth-vertical axis for 5 sec, sample-rate 100 Hz
```

```
# Experimental parameters
down = np.deg2rad(10)     # deg
duration = 10   # sec
rate = 100     # Hz

# Starting orientation of the camera
q_start = np.array([0, np.sin(down/2), 0])

# Movement of the platform
dt = 1./rate
t = np.arange(0, duration, dt)
omega = np.tile([0, 0, np.deg2rad(100)], (len(t), 1) )

# Orientation of the camera in space, during the rotation of
   the platform
# with a space-fixed ("sf") angular velocity "omega".
# Note that this one line does all the calculations required!
q_moving = quat.calc_quat(omega, q_start, rate, 'sf')

# Visualization
fig, axs = plt.subplots(1,2, figsize=(10,6))
axs[0].plot(t, quat.q_vector(q_moving))
axs[0].set(xlabel = 'Time [sec]',
       ylabel = 'Orientation [quat]',
       title = 'Rotation about g, with 100 deg/s')
axs[0].legend(['$q_1$', '$q_2$', '$q_3$'])
# Note that even for this simple arrangement,
   the camera-orientation
# in space is difficult to visualize!!

# And back from orientation to velocity
omega_recalc = quat.calc_angvel(q_moving, rate=rate)
axs[1].plot(t, np.rad2deg(omega_recalc))
axs[1].set(xlabel = 'Time [sec]',
       ylabel = 'Angular Velocity [deg/s]',
       title = 'Back-transformation')
axs[1].legend(['$\omega_x$', '$\omega_y$', '$\omega_z$'])

# Save to file
out_file = 'C5_examples_vel.png'
plt.savefig(out_file, dpi=200)
print('Image from "C5_examples_vel.py" saved to {0}.'
   .format(out_file))
plt.show()
```

C.7 Data Analysis of Movement Recordings

C.7.1 Analysis of Marker Recordings

The module `markers` in the package `scikit-kinematics` provides the following functionality:

markers.analyze_3Dmarkers Kinematic analysis of video-based recordings of 3-D markers.
markers.find_trajectory Calculation of point trajectory, from initial position + sensor position/orientation.

C.7.2 Analysis of Inertial-Sensor Recordings

The module `imus` in the package `scikit-kinematics` implements classes for IMU data from different producers (at the time of writing *XSens, x-IO, YEI, polulu*). These are based on a base-class `IMU_Base`. (In addition, data can also be entered directly, with the option `manual`.)

The advantage of this approach is that it allows to write code that is independent of the IMU sensor. All IMUs provide linear acceleration and angular velocities, and most of them also the direction of the local magnetic field. The specifics of each sensor are hidden in the sensor object (specifically, in the `get_data` method, which has to be implemented once for each sensor-type). Initialization of a sensor object includes a number of activities:

- Reading in the data.
- Making acceleration, angular_velocity etc. accessible in a sensor-independent way.
- Calculating `duration` and `totalSamples`.
- Calculating the position `pos`, with the method `calc_position`.
- Calculating orientation (expressed as `quat`), with the method specified in `q_type`. (See below the description of the function `_calc_orientation`, for the different options of `q_type`.)

The code sample below provides an example of how to use the `imus` module. It provides the following functionality:

imus.analytical Calculate orientation and position analytically, from angular velocity and linear acceleration.
imus.kalman Calculate orientation from IMU data using an *Extended Kalman Filter* as described by (Yun and Bachmann 2006).
imus.IMU_Base.calc_position Calculate the current position, assuming that the orientation is already known.
imus.IMU_Base.get_data Abstract method, to retrieve *rate*, *acc*, *omega*, *mag* from the input source. Overwritten with the specific method for each sensor type.

imus.IMU_Base.set_qtype Sets *q_type*, and automatically performs the relevant calculations. (Same options as `_calc_orientation`.)

imus.IMU_Base._calc_orientation Calculate the current orientation, using one of the following options:

- `analytical` ... quaternion integration of angular velocity
- `kalman` ... quaternion Kalman filter (Yun and Bachmann 2006)
- `madgwick` ... gradient descent method, efficient (Madgwick et al. 2011)
- `mahony` .... formula from Mahony, as implemented by Madgwick

The last two options for orientation calculation require the following classes for sensor integration, which are also provided:

imus.MahonyAHRS Madgwicks implementation of Mayhonys "AHRS" (attitude and heading reference system) algorithm.

imus.MadgwickAHRS Madgwicks gradient descent filter.

python™ **Code**: `C6_examples_IMU.py`: Example of working with sensors. See Fig. C.5.

Listing : C6_examples_IMU.py

```
"""Example for working with data from an IMU """

# author: Thomas Haslwanter, date: Oct-2018

import numpy as np
import matplotlib.pyplot as plt
import seaborn as sns

from skinematics.sensors.xsens import XSens


# Read in the recorded data. Here a file from an XSens-system
    :
data_file = r'data_xsens.txt'

# The easiest way to specify the approximate orientation
    is by indicating
# the approximate direction the(x,y,z)-axes of the IMU
    are pointing at:
x = [1, 0, 0]
y = [0, 0, 1]
z = [0,-1, 0]
initial_orientation = np.column_stack((x,y,z))

initial_position = np.r_[0,0,0]
orientation_calculation = 'analytical' # Method for
    orientation calculation
```

```
# Reading in the data, and initializing the ''Sensor''
    object. In this step also
# the orientation is calculated.
# To read in data from a different sensor, the corresponding
    class has to be
# imported from skinematics.sensors.
my_imu = XSens(in_file=data_file,
 q_type=orientation_calculation,
       R_init=initial_orientation,
       pos_init=initial_position)

# Example 1: extract the raw gyroscope data

gyr = my_imu.omega
time = np.arange(my_imu.totalSamples)/my_imu.rate

# Set the graphics parameters
sns.set_context('poster')
sns.set_style('ticks')

# Plot it in the left figure
fig, axs = plt.subplots(1,2, figsize=[18,8])
lines = axs[0].plot(time, gyr)
axs[0].set(title='XSens-data',
           xlabel='Time [s]',
           ylabel = 'Angular Velocity [rad/s]')
axs[0].legend(lines, ('x', 'y', 'z'))

# Example 2: extract the vector from the orientation
    quaternion,
#           which was calculated using an analytical procedure

q_simple = my_imu.quat[:,1:]

# Plot it in the right figure
lines = axs[1].plot(time, q_simple, label='analytical_')
axs[1].set(title='3D orientation',
           xlabel='Time [s]',
           ylabel = 'Quaternions')

# Example 3: calculate the orientation, using an extended
    Kalman filter
# Note that the orientation is automatically re-calculated
    when the
# "q_type" of the Sensor object is changed!

my_imu.set_qtype('kalman')
                                #executes the Kalman-filter
q_Kalman = my_imu.quat[:,1:]
```

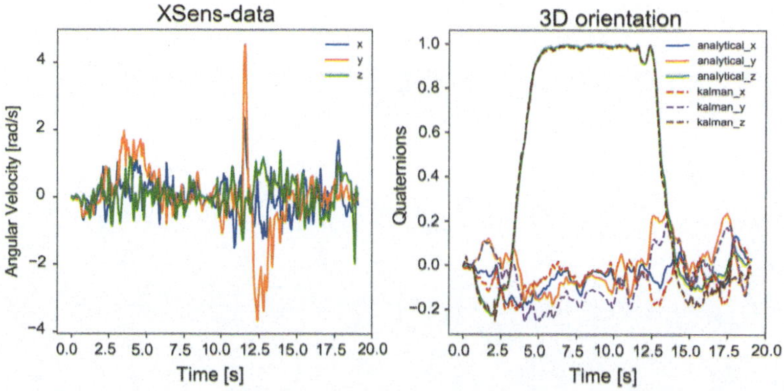

Fig. C.5 **Left**: angular velocity. **Right**: orientation from a sample movement recording with an XSens-sensor. The solid lines represent quaternions calculated analytically, and the dashed lines those from a Kalman filter. Note that the quaternion values, multiplied with 100, give approximately the orientation in degrees

```
# Superpose the lines on the right figure
lines.extend(axs[1].plot(time, q_Kalman,'--',
    label='kalman_'))

# Add the direction-info to each line label
dirs = ['x', 'y', 'z']
for ii, line in enumerate(lines):
     line.set_label(line.get_label() + dirs[ii%3])

plt.legend(loc='upper right')

# Save the figure
out_file = 'orientations.png'
plt.savefig(out_file, dpi=200)
print('Image saved to {0}'.format(out_file))

plt.show()
```

Appendix D
Human Movement Recordings—Practical Tips

D.1 Movement Recordings with a Marker-Based System

The goal of human movement recordings is typically to find the bone pose and the joint kinematics. This can be achieved with an initial calibration, which determines of the location of distinct anatomical landmarks with respect to the markers attached to the body. Different experimental approaches can be used for that (Camomilla and Vannozzi, 2018; Selbie and Brown, 2018. Once that information is available it is possible to reconstruct the exact limb movement from the measured marker positions.

The simplified paradigm described below describes how this can be achieved with three markers attached to the body, and an additional marker to indicate distinct anatomical landmarks:

1. Choose a space coordinate system (SCS).
2. Find position and orientation of the Optical Recording System (ORS) relative to SCS.
3. Place markers on the body.
4. Check visibility of markers during whole movement.
5. Find position and orientation of markers with respect to anatomical landmarks.
6. Bring subject in reference position and reference orientation.
7. Record movement.

ad (1) Choosing a Space Coordinate System

The first step in any movement recording is to establish a space-fixed coordinate system (SCS), and requires the positions of three or more points in space. A natural choice is a system where three points in the horizontal plane determine the direction of $\mathbf{n_x}$ and $\mathbf{n_y}$, and direction of gravity determines the $\mathbf{z}$-axis:

- $\mathbf{p_0}$: Origin of chosen inertial system, in ORS coordinates.
- $\mathbf{p_1}$: (Exact) positive direction of x-axis, as the direction from $\mathbf{p_0}$ to $\mathbf{p_1}$.
- $\mathbf{p_2}$: (Approximate) positive direction of y-axis. (The exact direction will be given by a Gram–Schmidt orthogonalization.)

T. Haslwanter, *3D Kinematics*, https://doi.org/10.1007/978-3-319-75277-8

ad (2) Finding Position and Orientation of ORS

Using these three points as inputs, a Gram–Schmidt orthogonalization [(Eq. (6.4)] provides the rotation matrix **R** describing the orientation of the SCS in ORS coordinates.

ad (3) Attaching Markers to the Body

Artifacts in the measured marker position can be caused by skin movement, and by so-called "soft tissue artifacts". The first one can be avoided in principle only by using bone pins, which are screwed into a human bone (Reinschmidt et al. 1997a, b). However, bone pins are—understandably—not only unpopular with most subjects, they are also difficult to justify for most ethics committees. Alternatively, movements of the markers with respect to the bones can be minimized by tightly fixating the markers to the underlying body part.

The second class of artifacts, soft tissue artifacts, can be caused by movement of muscles as well as by fat tissues. Those artifacts can be reduced by selecting body parts with the smallest amount of muscle mass and soft tissue (e.g., the wrist).

ad (4) Checking Marker Visibility

Markers that may be temporarily hidden during movement recordings can cause significant problems for 3-D movement recordings. Markers can be hidden by cables, by clothes, or by body parts.

In principle, there are four ways to minimize these problems:

- Using a nonoptical recording system.
- Multiple cameras and/or multiple markers.
- Careful planning and checking of the intended movements.

Most magnetic field systems, such as the *Aurora* or the *trakSTAR* system (both by *Ascension Technology Corporation*, now an affiliate of *Northern Digital Inc., NDI*), already provide position and velocity in space, and thereby remove many of the experimental restriction which can occur when using marker-based systems. Also IMUs, which are covered in the next chapter, do not require a direct line-of-sight to the recording system. However, they come at the cost of lacking information about absolute position (and for IMUs without magnetometers lacking information about absolute orientation in space).

By using multiple cameras, such as the *Vicon* system in Fig. 2.1, the problem of temporarily hidden markers is reduced. And multiple markers, which are mounted such that in principle always at least three of them are visible, can also compensate for markers that are temporarily hidden. However, in return, the analysis algorithms can become more complex.

The last solution suggested in the list above, "checking the intended movements", can often be implemented simply by standing behind the recording equipment, and asking the subject to perform the movement required for the experimental paradigm. Visibility of all markers during this simple visual check ensures that the experiment will most likely provide all the data required for a successful analysis.

ad (5) Recording Anatomical Landmarks

Many applications are not (only) interested in the orientation of the individual body parts, but instead focus on joint locations. Visualizations of human movements are a prime example of such applications, as are biomechanical investigations. Denoting *one* location for a joint is complicated by the fact that (i) it is impossible to place a marker inside a joint, and (ii) some joints execute highly complicated movements, for example, the combined rotation–translation of knee joints.

The most accurate information that can be achieved with ORSs is by determining the location of one or more anatomical landmarks, in a body orientation where also the three markers of the corresponding body part are visible. In the example in Fig. D.1, markers attached to the lower arm can record its position and orientation, and the location of the *inside* and the *outside* of the wrist can be recorded with an additional sensor, while the lower arm markers also have to be visible. Figure D.2 shows experimental data from such a recording in our laboratories. The corresponding software for the data analysis is provided in `scikit-kinematics`, in the module `markers`.

For example, for the accurate determination of the wrist position with respect to the lower arm markers, one can proceed as follows:

- While three markers are visible, hide the fourth marker with one hand, and position it at the inside of the wrist.
- Reveal that marker for a second or so, then hide it again, and place it on the outside of the wrist.
- Reveal that marker again for a second, and then hide it, before terminating the recording.

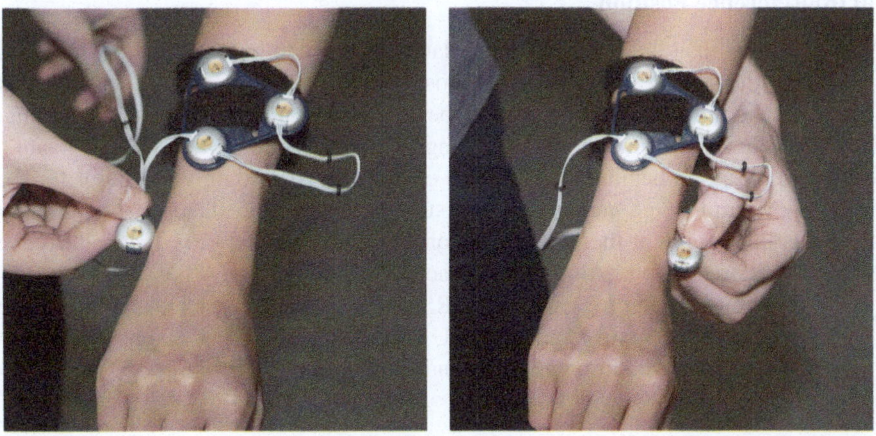

Fig. D.1 To find the position of the wrist with respect to the three markers mounted on the lower arm, the outer (left) and inner (right) edge of the wrist can be indicated with an additional marker

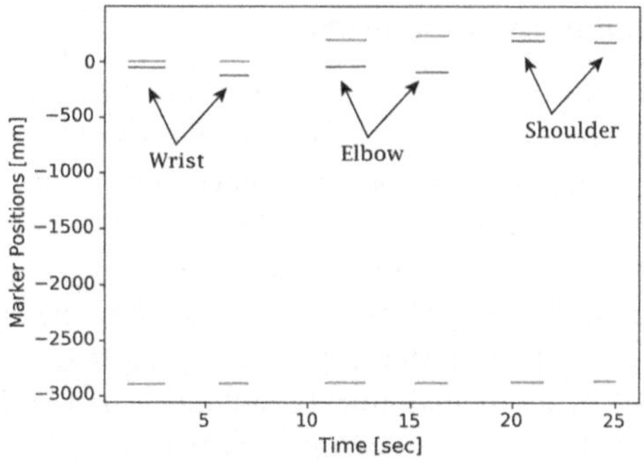

Fig. D.2 To identify the center of a joint, a calibration marker can be held to the inner and outer edges of that joint. Figure D.1 shows how this can be done for the wrist. In the recording shown in this Figure, this was done for the wrist, elbow, and shoulder, providing six calibration points. From these, the center of the wrist, the elbow, and the shoulder relative to the markers on the lower or upper arm can be determined

This sequence can be repeated for each joint of interest.

Figure D.2 shows raw data for such a recording, for that fourth "location"-marker. With an automated analysis procedure, or with interaction by the user, one point can be selected for the calculation of the position of the anatomical landmark.

From these data, the vector from the three limb markers to the anatomical landmark can be calculated, applying Eqs. (6.2) and (6.9), replacing the "COM" with the anatomical landmark.

Note: In practice, this procedure is often substituted by placing markers at well defined anatomical positions, from which the joint positions can then be calculated.

ad (6) Reference Position

Many artifacts that can complicate the data analysis can be eliminated or reduced by placing the subject in the reference position/orientation as often as possible. At the very least, this should be done at the beginning and at the end of each recording. For example, for eye movement recordings, the subject could be asked to fixate a known target at the beginning and at the end of each recording session. Or for the recording of arm or leg movements, the subject could try to stand or sit stationary at the beginning and at the end of a recording, with the body in exactly the same location and orientation. (Although in practice this is not always possible.)

For movement recordings with an ORS, such calibration checks can allow the detection of a marker slip. And for IMU recordings (see next chapter), they can provide a possibility to eliminate drifts and/or offsets.

ad (7) Movement Recording

Having gone through all the procedures described above increases the chances to obtain reliable, high-quality recordings of the movement pattern under investigation.

D.2 Movement Recordings with a Sensor-Based System (IMU)

The following steps indicate important aspects of experimental paradigms for movement recordings with inertial measurement units (IMUs):

1. Choose the Space Coordinate System (SCS).
2. Bring the subject in a static starting position.
3. Attach the IMU(s) to the body.
4. Document the orientation of the IMU(s).
5. Record the movement.
6. Bring subject again in the static starting position.

ad (1) Choosing a Coordinate System

This is important in order to orient the subject in step (2).

ad (2) Reference/Starting Position

In many applications, the mid-sagittal plane of the subject is parallel to the x-z-plane of the SCS.

ad (3) Attaching IMU to Body

The raw movement data are significantly easier to interpret if the IMU axes are as parallel to the axes of the SCS as possible. (See also Sect. 4.5.2, which describes the determination of the sensor orientation when the subject is stationary.)

ad (4) Documenting the Starting Position and the IMU Orientation

Given the easy availability of smartphones with cameras and/or digital cameras, it is very advisable to take pictures to document (i) the orientation of the IMU(s) on the body, and (ii) the starting position of the subject. If those two pieces of information are missing, IMU data are very hard to interpret!

ad (5) Movement Recording with IMU

- The subject should be stationary at the beginning of the recording.
- The recording sessions should be kept short, since most IMUs show a significant amount of drift.
- At the end of the recording, the subject should be stationary again, and if possible in exactly the same position as at the beginning of the recording. If this is the case, drift artifacts in the IMU signals can be eliminated in the data analysis.

Appendix E
Exercise Solutions

E.1 Warm-Up Exercises

Solution Exercise 1.1: A Simple Linear Movement
Integration of a sine function gives

$$acc(t) = amp * \sin(\omega t) \tag{E.1}$$

$$vel(t) = -\frac{amp}{\omega} * \cos(\omega t) + c\,. \tag{E.2}$$

With the initial condition $vel(t = 0) = v_0$ one gets

$$v_0 = -\frac{amp}{\omega} + c$$

$$c = \frac{amp}{\omega} + v_0\,, \tag{E.3}$$

which gives the velocity

$$vel(t) = \frac{amp}{\omega} * (1 - \cos(\omega t)) + v_0\,. \tag{E.4}$$

Another integration and application of the second initial condition $pos(t = 0) = p_0$ lead to the final solution

$$pos(t) = -\frac{amp}{\omega^2} * \sin(\omega t) + \left(\frac{amp}{\omega} + v_0\right) * t + p_0\,. \tag{E.5}$$

python **Code**: `Ex_1_1_Solution.py`: Example of working with quaternions.

T. Haslwanter, *3D Kinematics*, https://doi.org/10.1007/978-3-319-75277-8

Listing : Ex_1_1) 1-D Accelerometer - analytical & numerical solution.

```
"""Exercise 1.1: Solve equations of motion for sinusoidally
    moving accelerometer."""

# author: Thomas Haslwanter, date: Dec-2017

# Load the required packages
import numpy as np
import matplotlib.pyplot as plt

# Set up the parameters
duration, rate = 10, 50 # for the plot
freq, amp = 2, 5       # for the sine-wave
v0, p0 = 0, 15         # initial conditions

# Calculate derived values
omega = 2*np.pi*freq
dt = 1/rate
time = np.arange(0, duration, dt)
acc = amp * np.sin(omega*time)

# In the following, I put each task in a function, to
    facilitate readability

def analytical():
    """Analytical solution of a sinusoidal acceleration

    Returns
    -------
    axs : list
        Axes handles to the three plots.
    """

    # Analytical solution
    vel = amp/omega * (1-np.cos(omega*time)) + v0
    pos = -amp/omega**2 * np.sin(omega*time) + (amp/omega + v0
        )*time + p0

    # Plot the data
    fig, axs = plt.subplots(3,1,sharex=True)

    axs[0].plot(time, acc)
    axs[0].set_ylabel('Acceleration')
    axs[0].margins(0)

    axs[1].plot(time, vel)
    axs[1].set_ylabel('Velocity')
    axs[1].margins(0)

    axs[2].plot(time, pos, label='analytical')
```

```
    axs[2].set_ylabel('Position')
    axs[2].set_xlabel('Time [sec]')
    axs[2].margins(0)

    return axs

def simple_integration(axs):
    """Numerical integration of the movement equations

    Paramters:
    ----------
    axs : list
        Axes handles to the three plots produced by "
            analytical".
    """

    # Initial conditions
    vel, pos = [v0], [p0]

    # Numerical solution
    for ii in range(len(time)-1):
        vel.append(vel[-1] + acc[ii]*dt)
        pos.append(pos[-1] + vel[-1]*dt) # Euler-Cromer method

    # Superpose the lines on the previous plots
    axs[1].plot(time, vel)
    axs[2].plot(time, pos, label='numerical')
    plt.legend()

    plt.show()

if __name__ == '__main__':
    axs_out = analytical()
    simple_integration(axs_out)

    input('Done') # Without this line Python closes
        immedeatly after running
```

Solution Exercise 1.2: Find the Cat

To find the position of the cat (center between the eyes) in the figure, one has to

- Determine the location of the origin (big white circle in Fig. E.1).
- Fixate the orientation of the chosen coordinate system (*horizontal/vertical*).
- Measure the vertical/horizontal location (white lines in Fig. E.1), in pixel.
- Measure the size of the Ikea-shelf (yellow bar in Fig. E.1), in pixel.
- Convert the measured dimensions into cm.

With $ver = 258\,px$, $hor = 61\,px$, $ikea = 226\,px$, and the information that the shelf is 124 cm high, we get

Fig. E.1 The cat is 142 cm up, and 33 cm to the right

$$up = ver * \frac{124}{ikea} = 142\,\text{cm} \tag{E.6}$$

$$right = hor * \frac{124}{ikea} = 33\,\text{cm}. \tag{E.7}$$

The calculations assume that the projection into the 2-D image plane is a parallel projection, i.e., that the distance d to the camera is much larger than the focal length f. They also assume that the x-pixels and the y-pixels are symmetrical.

Solution Exercise 1.3: A Simple Pendulum

python™ **Code**: `Ex_1_3_Solution.py`: Example of working with quaternions (Fig. E.2).

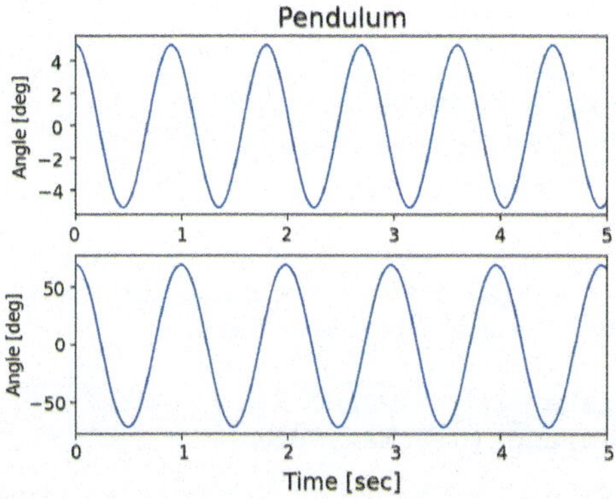

Fig. E.2 Pendulum trajectory, for initial deflection of 5° (top) and of 70° (bottom)

Listing : Ex_1_3) Movement trajectory for a pendulum, for small and large deviations.

```
"""Exercise 1.3: Simulation of a "simple" pendulum. """

# Author: Thomas Haslwanter, Date: April-2017

# First, import the required packages
import numpy as np
import matplotlib.pyplot as plt
import numpy as np
from scipy import constants
import os

# Define constants
g = constants.g   # gravity, [m/s^2]

def calculate_trajectory(phi_0, length_pendulum=0.2, out_file
    ='test.dat'):
    """Simulate a pendulum in 2-D
    The equation of motion is:
        \frac{d^2 \phi}{dt^2} = -\frac{g}{l} \sin(phi)
    Also, writes the data to an out-file.

    Parameters
    ----------
    phi_0 : float
        starting angle [deg].
    length_pendulum : float
        length of the pendulum [m]
        The default is 0.2 m
    out_file : string
        name of the out_file

    Returns
```

```
    -------
    time : ndarray
        Time samples [s]
    angle : ndarray
        Pendulum angle [deg]
    """

    tMax = 5     # duration of simulation [sec]
    rate = 1000 # sampling rate [Hz]

    dt = 1 / rate
    time = np.arange(0, tMax, dt)

    def acc_func(alpha):
        """ Differential equation of motion

        Parameters:
        -----------
        alpha : float
            Starting angle [deg]

        Returns:
        --------
        time : ndarray
            Time values for the movement of the pendulum.
        phi : ndarray
            Pendulum angle [deg]
        """
        acc = -g/length_pendulum * np.sin(alpha)
        return acc

    # Initial conditions
    phi = [np.deg2rad(phi_0)] # rad
    omega = [0] # rad/s

    # Numerical integration with the Euler-Cromer method,
        which is more stable
    for ii in range(len(time) - 1):
        phi.append(phi[-1] + omega[-1]*dt)
        omega.append(omega[-1] + acc_func(phi[-1])*dt)

    return time, np.rad2deg(phi)

if __name__ == '__main__':
    '''Main part'''

    pendulum = 0.20 # [m]
    'Call the function that calculates the trajectory, and
        generate a plot'
```

```
# For multiple plots, I clearly prefer the object oriented plotting style
# More info on that in the book "Introduction to Statistics with Python",
# in Chapter 3
fig, axs = plt.subplots(2,1)

# Starting position: 5 deg
time, angle = calculate_trajectory(5)
axs[0].set_xlim([0, np.max(time)])
axs[0].plot(time, angle)
axs[0].set_ylabel('Angle [deg]')
axs[0].set_title('Pendulum')

# Starting position: 70 deg
time, angle = calculate_trajectory(70)
axs[1].plot(time, angle)
axs[1].set_xlim([0, np.max(time)])
axs[1].set_ylabel('Angle [deg]')
axs[1].set_xlabel('Time [sec]')

# Save the figure, and show the user the corresponding info
outFile = 'pendulum.png'
plt.savefig(outFile, dpi=200)
print('out_dir: {0}'.format(os.path.abspath('.')))
print('Image saved to {0}'.format(outFile))
plt.show()
```

E.2 Rotation Matrices

Solution Exercise 3.1: CT Scanner

For the scanner shown in Fig. 3.19, the full rotation matrix is given by

$$\mathbf{R} = \mathbf{R}_z(\alpha) \cdot \mathbf{R}_x(\beta) \cdot \mathbf{R}_y(\gamma)$$

The task can be rephrased as: find the angles α, β, and γ which bring the scanner-fixed y-axis to align with $\overrightarrow{bullet_1}$, and the scanner-fixed z-axis with $\vec{n} = \overrightarrow{bullet_1} \times \overrightarrow{bullet_2}$. Care has to be taken, because the solution involves angles >90°.

python™ **Code**: `Ex_3_1_Solution.py`: Finding the orientation of the elements of a CT scanner.

Listing : Ex_3_1) CT-scanner.

```
""" Exercise 3.1: Orientation of the elements of a modern CT
    scanner.

"""

# Author: Thomas Haslwanter, Date: Dec-2017
import numpy as np
import os

from skinematics.rotmat import R, R_s
from skinematics.vector import normalize
from collections import namedtuple


def find_CT_orientation():
    '''Find the angles to bring an "Axiom Artis dTC" into a
        desired orientation.'''

    # Calculate R_total
    R_total = R_s(2, 'alpha')*R_s(0, 'beta')*R_s(1, 'gamma')

    # Use pprint, which gives a nicer display for the matrix
    import pprint
    pprint.pprint(R_total)

    # Find the desired orientation of the CT-scanner
    bullet_1 = np.array([5,2,2])
    bullet_2 = np.array([5,-2,-4])
    n = np.cross(bullet_1, bullet_2)

    ct = np.nan * np.ones((3,3))
    ct[:,1] = normalize(bullet_1)
    ct[:,2] = normalize(n)
    ct[:,0] = np.cross(ct[:,1], ct[:,2])

    print('Desired Orientation (ct):')
    print(ct)

    # Calculate the angle of each CT-element
    beta = np.arcsin( ct[2,1] )
    gamma = np.arcsin( -ct[2,0]/np.cos(beta) )

    # next I assume that -pi/2 < gamma < pi/2:
    if np.sign(ct[2,2]) < 0:
        gamma = np.pi - gamma

    # display output between +/- pi:
    if gamma > np.pi:
        gamma -= 2*np.pi
```

```
    alpha = np.arcsin( -ct[0,1]/np.cos(beta) )

    alpha_deg = np.rad2deg( alpha )
    beta_deg = np.rad2deg( beta )
    gamma_deg = np.rad2deg( gamma )

    # Check if the calculated orientation equals the desired
        orientation
    print('Calculated Orientation (R):')
    rot_mat = R(2, alpha_deg) @ R(0, beta_deg) @ R(1,
        gamma_deg)
    print(rot_mat)

    return (alpha_deg, beta_deg, gamma_deg)

if __name__=='__main__':

    (alpha, beta, gamma) = find_CT_orientation()
    print('alpha/beta/gamma = {0:5.1f} / {1:5.1f} / {2:5.1f}
            [deg]\n'.format(alpha, beta, gamma))
```

Solution Exercise 3.2: Combining Rotation and Translation

python **Code**: `Ex_3_2_Solution.py`: Combining rotation and translation.

Listing : Ex_3_2) Combining rotation and translation.

```
""" Exercise 3.1: Calculate the trajectory of an observed
    particle, when location and orientation of the observer
    are changing.
The velocities correspond to particle velocities in a central
     potential.

"""

# Author: Thomas Haslwanter, Date: Nov-2017

import os
from scipy import signal
import skinematics as skin
import numpy as np
import matplotlib.pyplot as plt

def rotate_and_shift(shift=[0,0,0], rotation=0):
    """Get data, and rotate and shift the camera location

    Parameters
    ----------
    shift : ndarray, shape (3,)
```

```
        Camera translation [Same units as position recordings]
    rotation : float
        Camera rotation [deg]
    """

    # Get the data
    inFile = 'planet_trajectory_2d.txt'
    data = np.loadtxt(inFile)

    # Calculate the 3D trajectory
    zData = -data[:,1]*np.tan(30*np.pi/180.)
    data3D = np.column_stack( (data,zData) )
    data3D[:,2] -= 200

    # Calculate and plot the 3-D velocity
    vel3D = signal.savgol_filter(data3D, window_length=61,
        polyorder=3, deriv=1, delta=0.1, axis=0)

    # Show the data
    plt.plot(vel3D[:,0], vel3D[:,1])
    plt.axis('equal')
    plt.title('x/y Velocity')
    plt.xlabel('x')
    plt.ylabel('y')
    plt.show()

    # Just to show how to elegantly create 3 subplots
    fig, axs = plt.subplots(3,1, sharex=True)
    for ii in range(3):
        axs[ii].plot(vel3D[:,ii])
        axs[ii].set_ylabel('axis_{0}'.format(ii))
    axs[ii].set_xlabel('Time [pts]')
    axs[0].set_title('Velocities')
    plt.show()

    # Shift the location of the camera, using the numpy
        feature of
    # "broadcasting" to subtract a 3-vector from an Nx3-matrix
        .
    data_shifted = data3D - np.array(shift)

    # Rotate the orientation of the camera
    data_shiftRot = (skin.rotmat.R(0, rotation) @ data_shifted
        .T).T

    # Plot the shifted and rotated trajectory
    outFile = 'shifted_rotated.png'

    plt.plot(data_shiftRot[:,0], data_shiftRot[:,1])
    plt.axhline(0, linestyle='dotted')
    plt.axvline(0, linestyle='dotted')
```

```
    plt.xlabel('x')
    plt.ylabel('y')
    plt.title('Shifted & Rotated')

    plt.savefig(outFile, dpi=200)
    print('Data saved to {0}'.format(outFile))
    plt.show()

if __name__=='__main__':
    rotate_and_shift([0, 100, -50], 34)
```

E.3 Analysis of 3-D Movement Recordings

Solution Exercise 6.1: An (Almost) Simple Rotation

For a fixed-axis rotation about a coordinate axis, we can find the orientation by simply integrating the angular velocity. A 45° rotation about the vertical axis therefore gives

$$\mathbf{R}_z(\theta = 45^\circ) = \begin{bmatrix} 0.71 & -0.71 & 0 \\ 0.71 & 0.71 & 0 \\ 0 & 0 & 1 \end{bmatrix}.$$

The answer to the second question requires the correct sequence of two rotations (Sect. 3.4):

$$\mathbf{R}_{combined} = \mathbf{R}_z(\theta = 45) \cdot \mathbf{R}_y(\phi = 30^\circ) = \begin{bmatrix} 0.61 & -0.71 & 0.35 \\ 0.61 & 0.71 & 0.35 \\ -0.5 & 0 & 0.87 \end{bmatrix}$$

Solution Exercise 6.2 Pendulum

python™ **Code**: `Ex_6_2_Solution.py`: Simulating the movement and the forces measured by an IMU on a pendulum. See Fig. E.3.

Listing : Ex_6_2) Pendulum

```python
"""Exercise 6.1: Simulation of a "simple" pendulum.
This is not as simple as it looks. The signs are a bugger
(force vs acceleration, up vs down, orientation of coordinate
    system). Also, I was surprised how quickly one gets
    artefacts in the reconstruction: already with a sample
    rate of 1 kHz, artefacts sneak in!

"""

# Author: Thomas Haslwanter, Date: Dec-2017

import numpy as np
import matplotlib.pyplot as plt
import numpy as np
from scipy import signal, constants, integrate
import pandas as pd

g = constants.g    # gravity, [m/s^2]


def generate_data(length_pendulum, out_file='test.dat'):
    """Simulate a pendulum in 2D, and write the data to an out
        -file.

    Parameters
    ----------
    length_pendulum : float
        Length of the pendulum [m]
    out_file : string
        Name of out-file

    Returns
    -------
    df : Pandas DataFrame
        Contains 'time', 'phi', 'omega', 'gifBHor', 'gifBVer'
    rate : float
        Sampling rate [Hz]

    Notes
    -----
    The equation of motion is

    \frac{d^2 \phi}{dt^2} = -\frac{g}{l} \sin(phi)
    """

    tMax = 2      # duration of simulation [sec]
    rate = 10000 # sampling rate [Hz]

    dt = 1 / rate
```

```
time = np.arange(0, tMax, dt)

def acc_func(alpha):
    """ differential equation of motion """
    acc = -g/length_pendulum * np.sin(alpha)
    return acc

# Memory allocation
omega = np.nan * np.ones(len(time))
phi   = np.nan * np.ones(len(time))

# Initial conditions
phi[0] = 0.1 # rad
omega[0] = 0 # rad/s
tStatic = 0.1 # initial static setup [sec]

# Numerical integration
for ii in range(len(time) - 1):
    phi[ii + 1] = phi[ii] + omega[ii] * dt
    if time[ii] < tStatic: # static initial condition
        omega[ii + 1] = 0
    else:
        # Euler-Cromer method, is more stable
        omega[ii + 1] = omega[ii] + acc_func(phi[ii + 1]) *
            dt

# Find the position, velocity, and acceleration
# The origin is at the center of the rotation
pos = length_pendulum * np.column_stack( (np.sin(phi),
    -np.cos(phi)) )

vel = signal.savgol_filter(pos, window_length=5, polyorder
    =3, deriv=1, delta=dt, axis=0)
acc = signal.savgol_filter(pos, window_length=5, polyorder
    =3, deriv=2, delta=dt, axis=0)

# Add gravity
accGravity = np.r_[0, g]
gifReSpace = accGravity + acc

# Transfer into a body-fixed system
gifReBody = np.array([rotate(gif, -angle) for (gif, angle)
     in zip(gifReSpace, phi)])

# Quickest way to write the "measured" data to a file,
    with headers
df = pd.DataFrame(np.column_stack((time, phi, omega,
    gifReBody)), columns=['time', 'phi', 'omega', 'gifBHor
    ', 'gifBVer'])
df.to_csv(out_file, sep='\t')
print('Data written to {0}'.format(out_file))
```

```
    return df, rate


def show_data(data, phi, pos, length):
    """Plots of the simulation, and comparison with original
        data

    Parameters
    ----------
    data : Pandas DataFrame
        Contains 'time', 'phi', 'omega', 'gifBHor', 'gifBVer'
    phi : ndarray, shape (N,)
        Angles of the reconstructed movement.
    pos : ndarray, shape (N,2)
        x/y-positions of the reconstructed movement.
    length : float
        Length of the pendulum.
    """

    fig, axs = plt.subplots(3, 1, figsize=(5,5))

    # Show Phi
    axs[0].plot(data['time'], phi)
    axs[0].xaxis.set_ticklabels([])
    axs[0].set_ylabel('Phi [rad]')

    # Show Omega
    axs[1].plot(data['time'], data['omega'])
    axs[1].xaxis.set_ticklabels([])
    axs[1].set_ylabel('Omega [rad/s]')

    # Show measured force
    axs[2].plot(data['time'], data[['gifBHor', 'gifBVer']])
    axs[2].set_xlabel('Time [sec]')
    axs[2].set_ylabel('GIF re Body [m/s^2]')
    axs[2].legend(('Hor', 'Ver'))
    plt.tight_layout()

    # x,y plot of the position
    fig2, axs2 = plt.subplots(1, 2)
    axs2[0].plot(pos[:, 0], pos[:, 1])
    axs2[0].set_title('Position plot')
    axs2[0].set_xlabel('X')
    axs2[0].set_ylabel('Y')

    axs2[1].plot(pos[:, 0])
    plt.hold(True)
    axs2[1].plot(length * np.sin(phi), 'r')

    plt.show()
```

```
def rotate(data, phi):
    """Rotate 2d data in column form

    Parameters
    ----------
    data : ndarray, shape (2,)
        x/y-data to be rotated.
    phi : float
        Angle of 2-D rotation.

    Returns
    -------
    rotated : ndarray, shape (2,)
        Rotated data.
    """

    Rmat = np.array([[np.cos(phi), -np.sin(phi)],
                     [np.sin(phi), np.cos(phi)]])
    rotated = (Rmat @ data.T).T

    return rotated


def reconstruct_movement(omega, gifMeasured, length, rate):
    """ From the measured data, reconstruct the movement

    Parameters
    ----------
    omega : ndarray, shape (N,)
        Angular velocity [rad/s]
    gifMeasured : ndarray, shape (N,2)
        Gravito-inertial force per unit mass [kg m/s^2]
    length : float
        Length of pendulum [m]

    Returns
    -------
    phi : ndarray, shape (N,)
        Angle of pendulum [rad]
    pos : ndarray, shape (N,2)
        x/y-position of pendulum [m]
    """

    dt = 1/rate

    # Initial orientation
    gif_t0 = gifMeasured[0]
    phi0 = np.arctan2(gif_t0[0], gif_t0[1])
```

```
    # Calculate phi, by integrating omega
    phi = integrate.cumtrapz(omega, dx=dt, initial=0) + phi0

    # From phi and the measured acceleration, get the movement
        acceleration
    accReSpace = np.array([rotate(gif, angle) - np.array([0, g
        ]) for (gif, angle) in zip(gifMeasured, phi)])

    # Position and Velocity through integration
    vel = np.nan * np.ones(accReSpace.shape)
    pos = np.nan * np.ones(accReSpace.shape)

    init = length * np.array([np.sin(phi[0]), -np.cos(phi[0])
        ])
    for ii in range(accReSpace.shape[1]):
        vel[:, ii] = integrate.cumtrapz(accReSpace[:, ii], dx=
            dt, initial=0)
        pos[:, ii] = integrate.cumtrapz(vel[:, ii], dx=dt,
            initial=0)
        pos[:, ii] += init[ii] # initial condition

    return phi, pos


if __name__ == '__main__':
    '''Main part'''

    pendulum = 0.20 # [m]
    sim_data, sample_rate = generate_data(pendulum)

    # Get the data: this is just to show how such data can be
        read in again
    # data = pd.io.parsers.read_table(outFile,
        skipinitialspace=True)

    # From the measured data, reconstruct the movement:
    # First, find the sampling rate from the time
    phi_calc, pos_calc = reconstruct_movement(omega=sim_data['
        omega'].values,
                                              gifMeasured=sim_data
                                                  [['gifBHor', '
                                                  gifBVer']].
                                                  values,
                                              length=pendulum,
                                              rate=sample_rate)

    # Show the results - this should be a smooth oscillation
    show_data(sim_data, phi_calc, pos_calc, pendulum)
```

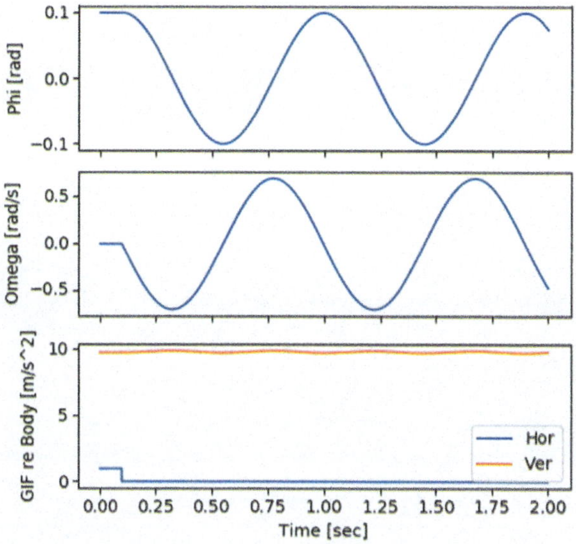

Fig. E.3 Pendulum angles and angular velocities, and the sensed gravito-inertial accelerations. Note in the bottom plot that when the pendulum is released, there is zero tangential acceleration ("Hor"). And in the "Ver" component, the oscillating gravitational component plus the centrifugal force induces a slight modulation

Appendix F
Glossary

Attitude Expression commonly used for *3-D orientation* in aeronautics.

Axis-angle representation Refers to the *Euler vector* (see below).

Cayley–Klein parameters The parameters α, β, γ, and δ, which provide a way to uniquely characterize the orientation of a solid body. The Cayley–Klein parameters are isomorphic to the group of *unit quaternions*. In terms of the Euler parameters e_i and the Pauli matrices σ_i, the Q-matrix

$$Q = \begin{bmatrix} \alpha & \beta \\ \gamma & \delta \end{bmatrix}$$

can be written as

$$Q = e_0 * I + j(e_1 * \sigma_1 + e_2 * \sigma_2 + e_3 * \sigma_3).$$

Euler's formula $e^{j\phi} = \cos\phi + j\sin\phi$.

Euler parameters The four parameters e_0, e_1, e_2, e_3 describing a finite rotation about an arbitrary axis. e_0 corresponds to the scalar part of a unit quaternion, and $\mathbf{e} = (e_1/e_2/e_3)$ to the vector part. Euler showed that these parameters define the rotation matrix via Eq. (4.15).

Euler vector Unique representation of a rotation, by a vector parallel to the axis of the rotation, and a length proportional to the magnitude of the rotation in radians. In other words, a quaternion vector, scaled to degree (see Sect. 4.2). Euler vectors have been mainly avoided in this book, because unlike quaternions and Gibbs vectors they cannot be used for calculations, e.g., for a combination of two rotations.

T. Haslwanter, *3D Kinematics*, https://doi.org/10.1007/978-3-319-75277-8

Euler–Rodrigues formula Describes the rotation of a vector in three dimensions. It is based on Rodrigues' rotation formula (see below), but uses a different parametrization. A rotation about the origin is represented by four real numbers, a, b, c, d such that $a^2 + b^2 + c^2 + d^2 = 1$. The parameter a may be called the *scalar* parameter, while $\boldsymbol{\omega} = b/c/d$ is the *vector* parameter. With this definition

$$\mathbf{x}' = \mathbf{x} + 2 * a * (\boldsymbol{\omega} \times \mathbf{x}) + 2 * (\boldsymbol{\omega} \times (\boldsymbol{\omega} \times \mathbf{x})) . \tag{F.1}$$

Euler rotation theorem There are many *Euler theorems*. In geometry, Euler's rotation theorem states that, in 3-D space, any displacement of a rigid body such that a point on the rigid body remains fixed, is equivalent to a single rotation about some axis that runs through the fixed point. It also means that the composition of two rotations is again a rotation. Therefore, the set of rotations has a group structure, known as the "rotation group".

Gibbs vector With q_0 the scalar part of a unit quaternion and $\mathbf{q}$ the vector part, a *Gibbs vector* $\mathbf{g}$ is defined as $\mathbf{g} = \frac{\mathbf{q}}{q_0}$.

Hamilton's identity The idea for quaternions occurred to Hamilton 1843 while he was walking along the Royal Canal on his way to a meeting of the Irish Academy. Hamilton was so pleased with his discovery that he scratched the fundamental formula of quaternion algebra

$$\tilde{i}^2 = \tilde{j}^2 = \tilde{k}^2 = \tilde{i}\tilde{j}\tilde{k} = -1, \tag{F.2}$$

into the stone of the Brougham bridge in Dublin, as he paused on it.

Heading Direction in which an aircraft or ship is heading.

Pauli matrices A set of three complex 2*x*2 matrices, which are Hermitian and unitary, and exponentiate to the *special unitary group* $SU(2)$:

$$\sigma_1 = \sigma_x = \begin{bmatrix} 0 & 1 \\ 1 & 0 \end{bmatrix}$$

$$\sigma_2 = \sigma_y = \begin{bmatrix} 0 & -i \\ i & 0 \end{bmatrix}$$

$$\sigma_3 = \sigma_z = \begin{bmatrix} 1 & 0 \\ 0 & -1 \end{bmatrix}.$$

The algebra generated by the three matrices $\sigma_1, \sigma_2, \sigma_3$ is isomorphic to the Clifford algebra of $\mathbf{R}^3$, called the "algebra of physical space". The three Pauli matrices of course go along with a fourth matrix, the unit matrix: $\begin{bmatrix} 1 & 0 \\ 0 & 1 \end{bmatrix}$. The exponential of the Pauli matrices shows their close relation to rotations:

$$e^{ja(\boldsymbol{n}\cdot\boldsymbol{\sigma})} = I\cos(a) + j(\boldsymbol{n}\cdot\boldsymbol{\sigma})\sin(a) \tag{F.3}$$

which is analogous to Euler's formula. **n** is a unit vector.

Quaternions Four-element algebra, allowing rotation and scaling of a quaternion by multiplying it with another quaternion. Includes *unit quaternions* as a subalgebra, which describes proper rotations. The $(\tilde{\boldsymbol{i}}, \tilde{\boldsymbol{j}}, \tilde{\boldsymbol{k}})$ quaternions behave like Pauli matrices multiplied by a factor j, where $j := \sqrt{-1}$. Specifically, one can verify that redefining $\tilde{\boldsymbol{i}} := j * \sigma_x$, $\tilde{\boldsymbol{j}} := j * \sigma_y$, and $\tilde{\boldsymbol{k}} := j * \sigma_z$ then once again we can write Hamilton's identities, namely $\tilde{\boldsymbol{i}}^2 = \tilde{\boldsymbol{j}}^2 = \tilde{\boldsymbol{k}}^2 = \tilde{\boldsymbol{i}}\tilde{\boldsymbol{j}}\tilde{\boldsymbol{k}} = -1$.

Rotation vectors This term is used by some authors for *Gibbs vectors*. Since this expression is sometimes found misleading ("Is a quaternion vector also a rotation vector?"), it has been avoided in this book.

Rodrigues parameters Components of Gibbs vector.

Rodrigues' rotation formula If **v** is a 3-D vector and **k** is a unit vector describing an axis of rotation about which **v** rotates by an angle θ, the *Rodrigues formula* says

$$\mathbf{v}_{\text{rot}} = \mathbf{v} * \cos(\theta) + (\mathbf{k} \times \mathbf{v}) * \sin(\theta) + \mathbf{k} * (\mathbf{k} \cdot \mathbf{v}) * (1 - \cos(\theta)\ . \tag{F.4}$$

Rotation group SO(3) The "special orthogonal group" (SO3) is the group of real-valued three-dimensional (3×3) orthogonal $(\mathbf{R}\mathbf{R}^T = \mathbf{R}^T\mathbf{R} = \mathbf{I})$ matrices with $det(R) = +1$. Those are the "rotation matrices".

Appendix G
Online Resources

G.1 Links Directly Related to this Book

3D_Kinematics Homepage of this book: Code Listings (Python & Matlab), Exercise Solutions, Errata.
https://github.com/thomas-haslwanter/3D_Kinematics

scikit-kinematics Source Code Python package accompanying this book.
https://github.com/thomas-haslwanter/scikit-kinematics

scikit-kinematics Documentation Python package accompanying this book.
http://work.thaslwanter.at/skinematics/html/

Matlab Kinematics Toolbox Matlab toolbox accompanying this book.
https://github.com/thomas-haslwanter/kinematics_toolbox.git

G.2 Other Links

Anaconda Windows/Linux/MacOS distribution for scientific work with Python.
https://www.anaconda.com/download/

Clifford Python package for Clifford algebra.
http://clifford.readthedocs.io

Clifford Algebra—Introduction Good starting point to delf a bit deeper into Clifford algebras.
https://www.av8n.com/physics/clifford-intro.htm

git Code versioning system.
https://git-scm.com/

T. Haslwanter, *3D Kinematics*, https://doi.org/10.1007/978-3-319-75277-8

git—Starting instructions Good, short and simple starting instructions to *git*, in many languages.
http://rogerdudler.github.io/git-guide/

github Online repository for open-source code.
https://github.com/

IPython Homepage for *IPython*, the package for interactive work with Python.
http://ipython.org/

IPython Tutorial *IPython* tutorial.
https://ipython.readthedocs.io

Madgwick Filter -1 C, C#, and Matlab implementation of the Madgwick filter.
http://x-io.co.uk/open-source-imu-and-ahrs-algorithms/

Madgwick Filter -2 Python implementation of the Madgwick filter.
http://work.thaslwanter.at/skinematics/html/imus.html

PyCharm A very good Python IDE.
https://www.jetbrains.com/pycharm/

Python Scientific Lecture Notes One of the best places to get started.
http://scipy-lectures.github.com

Quaternions—Introduction A good starting point for quaternions.
https://en.wikipedia.org/wiki/Quaternion

Quaternions—half-angle rule Explanation of the half-angle rule.
http://math.stackexchange.com/questions/302465/half-sine-and-half-cosine-quaternions

sympy Python package for symbolic computations.
http://www.sympy.org/en/index.html

TortoiseGit Windows shell interface to *git*.
https://tortoisegit.org/

Wing My favorite Python IDE.
https://wingware.com/

WinPython Windows distribution for scientific work with Python.
https://winpython.github.io/

References

Altmann, S. (1986). *Rotations, quaternions and double groups*. Oxford: Clarendon Press.

Bergamini, E., Ligorio, G., Summa, A., Vannozzi, G., Cappozzo, A., & Sabatini, A. M. (2014). Estimating orientation using magnetic and inertial sensors and different sensor fusion approaches: accuracy assessment in manual and locomotion tasks. *Sensors*, *14*(10), 18625–18649.

Brand, L. (1948). *Vector and tensor analysis*. New York: Wiley.

Camomilla V., C. A. & Vannozzi, G. (2018). *Three-Dimensional Reconstruction of the Human Skeleton in Motion*, chapter 1.2, (pp. 17–45). Springer, Cham.

Collewijn, H., Van der Steen, J., Ferman, L., & Jansen, T. C. (1985). Human ocular counterroll: assessment of static and dynamic properties from electromagnetic scleral coil recordings. *Experimental brain research*, *59*, 185–196.

Dai, J. S. (2015). Euler-rodrigues formula variations, quaternion conjugation and intrinsic connections. *Mechanism and Machine Theory*, *92*, 144–152.

Diebel, J. (2006). Representing Attitude: Euler Angles, Unit Quaternions, and Rotation Vectors. http://www.swarthmore.edu/NatSci/mzucker1/e27/diebel2006attitude.pdf. Accessed: 2017-12-04.

Euler, L. (1775). Formulae generales pro translatione quacunque corporum rigidorum. *Novi Comm.Acad.Sci.Imp.Petrop.*, 20, 189–207.

Fick, A. (1854). Die Bewegungen des menschlichen Augapfels. *Z Rat Med N F*, *4*, 109–128.

Filippeschi, A., Schmitz, N., Miezal, M., Bleser, G., Ruffaldi, E., & Stricker, D. (2017). Survey of motion tracking methods based on inertial sensors: A focus on upper limb human motion. *Sensors*, *17*(6), 1257.

Funda, J. & Paul, R. (1988). A comparison of transforms and quaternions in robotics. *IEEE Transactions on Robotics and Automation*, *7*, 886–891.

Girard, P. R. (1984). The quaternion group and modern physics. *Eur. J. Phys*, *5*, 25–32.

Goldstein, H. (1980). *Classical Mechanics* (Vol. 2). Reading, MA: Addison-Wesley.

Grassmann, H. G. (1844). *Ausdehnungslehre*. Leipzig: Teubner.

Hamilton, W. (1844). On quaternions, or on a new system of imaginaries in algebra. *Philosophical Magazine*, *25*(3), 489–495.

Harris, C. & Stephens, M. (1988). A combined corner and edge detector. In *Alvey vision conference*, volume 15 / 50 (pp. 10–5244).: Manchester, UK.

Haslwanter, T. (1995). Mathematics of three-dimensional eye rotations. *Vision research*, *35*(12), 1727–1739.

T. Haslwanter, *3D Kinematics*, https://doi.org/10.1007/978-3-319-75277-8

Haslwanter, T., Buchberger, M., Kaltofen, T., Hoerantner, R., & Priglinger, S. (2005). See++: A biomechanical model of the oculomotor plant. *Annals of the New York Academy of Sciences*, *1039*(1), 9–14.

Hassan, Y., & Canaan, R. (1991). Full-field bubbly flow velocity measurements using a multiframe particle tracking technique. *Experiments in Fluids*, *12*(1), 49–60.

Helmholtz, H. (1867). *Handbuch der physiologischen Optik*. Leipzig: Voss.

Hepp, K. (1990). *On listing's law. Commun Math Phys*, *132*, 285–292.

Hepp, K., Henn, V., Vilis, T., Cohen, B., & Goldberg, M. (1989). Brainstem regions related to saccade generation. In R. Wurtz (Ed.), *The neurobiology of saccadic eye movements* (pp. 105–212). Amsterdam: Elsevier.

Higgins, W. T. (1975). A comparison of complementary and Kalman filtering. *IEEE Transactions on Aerospace and Electronic Systems*, *3*, 321–325.

Jiménez, A. R., Seco, F., Prieto, J. C., & Guevara, J. (2010). Indoor pedestrian navigation using an ins/ekf framework for yaw drift reduction and a foot-mounted imu. In *Positioning Navigation and Communication (WPNC), 2010 7th Workshop on* (pp. 135–143).: IEEE.

Kar, A. (2010). Skeletal tracking using microsoft kinect. *Methodology*, *1*, 1–11.

Kindratenko, V. V. (2000). A survey of electromagnetic position tracker calibration techniques. Virtual Reality, 5(3), 169–182.

Kuipers, J. (1999). *Quaternions and Rotation Sequences*. Princeton University Press.

Leong, P., & Carlile, S. (1998). Methods for spherical data analysis and visualization. *Journal of neuroscience methods*, *80*, 191–200.

Ley, C. & Verdebout, T. (2017). *Modern Directional Statistics*. Chapman and Hall/CRC.

Madgwick, S. O. H., Harrison, A. J. L., & Vaidyanathan, A. (2011). Estimation of imu and marg orientation using a gradient descent algorithm. *IEEE International Conference on Rehabilitation Robotics : [proceedings], 2011*, 5975346.

Mahony, R., Hamel, T., & Pflimlin, J.-M. (2008). Nonlinear complementary filters on the special orthogonal group. *IEEE Transactions on automatic control*, *53*(5), 1203–1218.

Mardia, K. V. & Jupp, P. E. (1999). *Directional Statistics*. Wiley.

McConnell, J. J. (2005). *Computer graphics: theory into practice*. Jones & Bartlett Learning.

Miller, J., Rossi, E., Konishi, S., & Abramoff, M. (2003). Visualizing ocular tissue movement with little gold beads. *Investigative Ophthalmology & Visual Science*, *44*(13), 3123–3123.

Müller, B., Wolf, S., Brueggemann, G.-P., Deng, Z., Miller, F., & Selbie, W. S., Eds. (2018). *Handbook of Human Motion*. Springer International Publishing.

Reinschmidt, C., van Den Bogert, A. J., Murphy, N., Lundberg, A., & Nigg, B. M. (1997a). Tibiocalcaneal motion during running, measured with external and bone markers. *Clinical biomechanics (Bristol, Avon)*, 12, 8–16.

Reinschmidt, C., van den Bogert, A. J., Nigg, B. M., Lundberg, A., & Murphy, N. (1997b). Effect of skin movement on the analysis of skeletal knee joint motion during running. *Journal of biomechanics*, *30*, 729–732.

Robinson, D. (1963). A method of measuring eye movement using a scleral search coil in a magnetic field. *IEEE Trans. Biomed. Eng*, *10*, 137–145.

Rodrigues, O. (1840). Des lois gomtriques qui rgissent les dplacements d'un systme solide dans l'espace, et de la variation des coordonnes provenant de ses dplacements considrs indpendamment des causes qui peuvent les produire. *J de Mathmatiques Pures et Appliques*, *5*, 380–440.

Roetenberg, D., Slycke, P. J., & Veltink, P. H. (2007). Ambulatory position and orientation tracking fusing magnetic and inertial sensing. *IEEE Transactions on Biomedical Engineering*, *54*(5), 883–890.

Rooney, J. (1977). A survey of representation of spatial rotation about a fixed point. *Environment and Planning B*, *4*, 185–210.

Sabatini, A. M. (2006). Quaternion-based extended Kalman filter for determining orientation by inertial and magnetic sensing. *IEEE Transactions on Biomedical Engineering*, *53*(7), 1346–1356.

Savage, P. G. (2006). A unified mathematical framework for strapdown algorithm design. *Journal of Guidance, Control, and Dynamics*, *29*(2), 237–249.

Seel, T., Raisch, J., & Schauer, T. (2014). Imu-based joint angle measurement for gait analysis. *Sensors*, *14*(4), 6891–6909.
Selbie, W. S. & Brown, M. J. (2018). *3D Dynamic Pose Estimation from Marker-Based Optical Data*, chapter 2.1, (pp. 81–100). Springer, Cham.
Tweed, D., Cadera, W., & Vilis, T. (1990). Computing three-dimensional eye position quaternions and eye velocity from search coil signals. *Vision Res.*, *30*(1), 97–110.
Tweed, D., & Vilis, T. (1987). Implications of rotational kinematics for the oculomotor system in three dimensions. *J. Neurophysiol.*, *58*(4), 832–849.
van der Geest, J. N., & Frens, M. A. (2002). Recording eye movements with video-oculography and scleral search coils: a direct comparison of two methods. *Journal of neuroscience methods*, *114*, 185–195.
Westheimer, G. (1957). Kinematics of the eye. *J. Opt. Soc. Am.*, *47*, 967–974.
Wolpert, D. M., & Ghahramani, Z. (2000). Computational principles of movement neuroscience. *Nat Neurosci*, *3*(Suppl), 1212–1217.
Woodman, O. (2007). An introduction to inertial navigation. http://www.cl.cam.ac.uk/techreports/UCAM-CL-TR-696.html.
Wu, G., & Cavanagh, P. R. (1995). Isb recommendations for standardization in the reporting of kinematic data. *Journal of biomechanics*, *28*, 1257–1261.
Yun, X., & Bachmann, E. R. (2006). Design, implementation, and experimental results of a quaternion-based kalman filter for human body motion tracking. *IEEE transactions on Robotics*, *22*(6), 1216–1227.

Index

A
Accelerometers, linear, 17
Aerial gun, 48
Anaconda, 5
Anatomical landmarks, 157
Angle, between two vectors, 135
Angular velocity tensor, 79
Artifacts
  IMU, 19
  ORS, 156
Attitude, 29, 36
Axis-angle representation, 59

B
Banking, 36
Bayesian inference, 105
Bivector, 126
Bone pins, 16

C
Cardan angles, 45
Cardan joint, 45
Cartesian coordinate system, 29, 31
Center of Mass (COM), 87
Clifford algebra, 60, 122
Code versioning, 7
Commutativity, 29
Complementary filters, 111
Coriolis force, 20
Corner detection, 11
Correlation, 104
Covariance matrix, 103
Cross product, 115
CT scanner, 54

D
Dead reckoning, 100
Declination, 24
Denavit-Hartenberg transformation, 129
Direction Cosine Matrix (DCM), 31
Dispersion, 128
Distribution
  Gaussian, 100
  normal, 100
Dot product, 123

E
Elevation, 36
Equations of motion, 75
Euler angles, 45
  proper, 45
Euler–Rodrigues formula, 180
Euler's theorem, 30
Euler vector, 57, 59
Exterior product, 123

F
Faraday's law, 26
Fick angles, 41
Filter
  Kalman, 105
  Savitzky–Golay, 77
Finite difference approximations, 77
Force
  Coriolis, 20
  gravito-inertial (GIF), 17

G
Gait analysis, 45

T. Haslwanter, *3D Kinematics*, https://doi.org/10.1007/978-3-319-75277-8

Gaussian distribution, 100
 2-dimensional, 102
 combination of two, 101
Gaze, 41
Geometric algebra, 122
Geometric product, 123, 125
Gibbs vector, 66
 relation to quaternions, 66
Gimbal, 40
Gimbal lock, 45
Github, 8
Gradient descent, 112
Gram–Schmidt orthogonalization, 87, 118
Gravito-inertial force, 17
Gyroscope, 20

H
Half-angle rule, 81
Heading, 36
Homogeneous coordinates, 46

I
Image formation, 13
Image registration, 11
IMU
 artifacts, 19
 inertial measurement unit, 17
Inclination, 24
Induction, 25
Inertial sensor, 16
Inertial system, 29
Interest point, 11

K
Kalman filter, 105
 extended, 111
 unscented, 111
Kalman gain, 102
Kinect, 11

L
Lens
 equation, 15
 thin, 15
Listing's plane, 81
Lorentz force, 24

M
Magnetic flux, 25
Magnetometer, 17, 24
Markers, 11, 86
 visibility, 156
Matlab, 6
Matrix multiplication, 116
Movement, 75

N
Newton's laws
 third, 18
Newton's second law, 9, 75
Noncommutativity, 39
Normalization, of a vector, 136
Nutation, 43

O
Optical Recording System (ORS), 13
Optical sensors, 85
Optotrak, 5
Orientation, 40
 in space, 91
 of a plane, 137

P
Pendulum, 9, 97
Pitch, 36, 41
Polar coordinates, 32
Pose, 2
Position
 in space, 91
Precession, 43
Projection
 central, 15
 onto a vector, 136
 onto flat surface, 49
 parallel, 13
Projective coordinates, 46
Python, 5

Q
Quaternion, 57
 inverse, 61
 pure, 61
 relation to rotation matrix, 63
 rotation, 61
 scalar component, 60
 unit, 61
 vector component, 60

R
Reference orientation, 29
Revision control, 7
Right-hand rule, 36
Rodrigues parameters, 65
Roll, 36, 41
Rotation
  2-D, 30
  3-D, 34
  active, 40
  combined, 38
  in a plane, 30
  nested, 52
  of coordinate system, 37
  of object, 37
  passive, 40
  sequence, 40
  single axis, 30
Rotation matrix, 29
Rotation quaternion, 61
Rotor, 125

S
Scalar product, 115
Scikit-kinematics, 5
Search coils, 25
Sensor fusion, 99
Sensor integration, 99
Sequence
  Euler, 43
  Fick, 41
  Helmholtz, 42
  nautical, 41
Spatial transformation matrix, 47
Spherical statistics, 127
Standard deviation, 101
State prediction, 107

T
Tait–Bryan angles, 45
Trigonometry, 117
Trilateration, 13, 22

U
Unit quaternion, 61

V
Variance, 101
Vector decomposition, 118
Vector product, 115
Vector rotation, 137
Velocity, 75
  angular, 79
  linear, 76
Version control, 7

W
Wedge product, 123
WinPython, 5

Y
Yaw, 36, 41

The manufacturer's authorised representative in the EU is Springer Nature Customer Service Centre GmbH, Europaplatz 3, 69115 Heidelberg, Germany. If you have any concerns regarding our products, please contact ProductSafety@springernature.com

Printed and bound by CPI Group (UK) Ltd, Croydon, CR0 4YY
30/09/2026
02225229-0002